An Introduction to Chaos in Nonequilibrium Statistical Mechanics

This book is an introduction to the applications in nonequilibrium statistical mechanics of chaotic dynamics, and also to the use of techniques in statistical mechanics important for an understanding of the chaotic behaviour of fluid systems.

The fundamental concepts of dynamical systems theory are reviewed and simple examples are given. Advanced topics including SRB and Gibbs measures, unstable periodic orbit expansions, and applications to billiard systems, are then explained. The text emphasises the connections between transport coefficients, needed to describe macroscopic properties of fluid flows, and quantities, such as Lyapunov exponents and Kolmogorov–Sinai entropies, which describe the microscopic, chaotic behaviour of the fluid. Later chapters consider the roles of the expanding and contracting manifolds of hyperbolic dynamical systems and the large number of particles in macroscopic systems. Exercises, detailed references and suggestions for further reading are included.

This book will be of interest to graduate students and researchers, with a background in statistical mechanics, working in condensed matter physics, nonlinear science, theoretical physics, mathematics and theoretical chemistry.

JAY ROBERT DORFMAN attended the John Hopkins University, receiving a BA degree in chemistry in 1957, and a PhD degree in physics in 1961. Professor Dorfman then spent three years as a post-doctoral fellow at the Rockefeller University before moving to the University of Maryland as an Assistant Professor in the Institute for Fluid Dynamics and Applied Mathematics (now the Institute for Physical Science and Technology) and the Department of Physics and Astronomy. He was promoted to the rank of Professor in 1972. During the years 1983–1992 Professor Dorfman served as Director of the Institute for Physical Science and Technology, the Dean of the College of Computer Mathematical and Physical Sciences, the Vice President for Academic Affairs and Provost of the University of Maryland at College Park. Currently he is engaged in research on the relation between dynamical systems theory and non-equilibrium statistical mechanics.

CAMBRIDGE LECTURE NOTES IN PHYSICS 14

General Editors: P. Goddard, J. Yeomans

1. Clarke: The Analysis of Space-Time Singularities
3. Sciama: Modern Cosmology and the Dark Matter Problem
4. Veltman: Diagrammatica – The Path to Feynman Rules
5. Cardy: Scaling and Renormalization in Statistical Physics
6. Heusler: Black Hole Uniqueness Theorems
7. Coles and Ellis: Is the Universe Open or Closed?
8. Razumov and Saveliev: Lie Algebras, Geometry, and Toda-type Systems
9. Forshaw and Ross: Quantum Chromodynamics and the Pomeron
10. Jensen: Self-organised Criticality
11. Vanderzande: Lattice Models of Polymers
12. Esposito: Dirac Operators and Spectral Geometry

This book is dedicated to my wife Celia, to our children and grandchildren, and to the memory of our granddaughter Abby, whose laughter brought us joy, and whose courage gave us strength.

An Introduction to Chaos in Nonequilibrium Statistical Mechanics

J. R. DORFMAN

Institute for Physical Science and Technology
and
Department of Physics
University of Maryland
College Park, Maryland

PUBLISHED BY THE PRESS SYNDICATE OF THE UNIVERSITY OF CAMBRIDGE
The Pitt Building, Trumpington Street, Cambridge, United Kingdom

CAMBRIDGE UNIVERSITY PRESS
The Edinburgh Building, Cambridge CB2 2RU, UK
40 West 20th Street, New York, NY 10011–4211, USA
477 Williamstown Road, Port Melbourne, VIC 3207, Australia
Ruiz de Alarcón 13, 28014 Madrid, Spain
Dock House, The Waterfront, Cape Town 8001, South Africa

http://www.cambridge.org

First published 1999
Reprinted 2001

Typeset by the Author [CRC]

A catalogue record for this book is available from the British Library

Library of Congress Cataloguing in Publication data
Dorfman, J. Robert (Jay Robert), 1937–
An introduction to chaos in nonequilibrium statistical mechanics /
J. R. Dorfman
p. cm. – (Cambridge lecture notes in physics 14)
Includes bibliographical references and index.
ISBN 0 521 65589 7 (pbk.)
1. Statistical mechanics. 2. Chaotic behavior in systems.
I. Title. II. Series.
QC174.8.D67 1999
530.13′01′1857–dc21 98–50545 CIP

ISBN 0 521 65589 7 paperback

Transferred to digital printing 2003

Contents

Preface

v'Ha-aretz hayta tohu va'vohu...v'ruach Elohim merahefet al pnai ha-mayim.

Now the earth was unformed and void ... and the spirit of God hovered over the face of the waters.

Genesis, 1.2

This book began its life as a set of lecture notes based on a series of lectures given to fourth year students at the Institute for Theoretical Physics at the University of Utrecht during the spring semester of 1994. The course of lectures was entitled *From Molecular Chaos to Dynamical Chaos*. At the suggestion of Prof. Matthieu Ernst, two students in the class, Lucas Neevens van Baal and Iris Lafaille, took notes, edited them, and prepared a LaTeX manuscript that formed the basis for the lecture notes. The notes have undergone several revisions and many more corrective exercises to remove many errors that I inadvertently added to the original LaTeX file provided by Mr. van Baal and Ms. Lafaille. It is due to their hard work and desire to make the notes as clear as possible that the notes have made their reappearance as a book.

I would like to thank the students who attended the course in Utrecht and those who have used the lecture notes at the University of Maryland since then. One of the original students in Utrecht, Ramses van Zon, and some students at Maryland, Rainer Klages, Thomas Gilbert, Debabrata Panja, and Luis Nasser, have taken the subjects discussed here as Ph.D. theses topics, which, of course, is a pleasant outcome of any series of lectures to advanced students. I want to thank them as well as the many other students and colleagues who have attended my lectures for helping me understand and clarify many of the points presented here.

I would like to thank my colleagues at the Institute for Theoretical Physics, University of Utrecht, for their warm hospitality, their generous help, and their creatively critical attitude, which was always refreshing. It is a pleasure to thank two colleagues in particular, Matthieu Ernst and Henk van Beijeren, for all of their help, support, and advice, and for many fruitful scientific collaborations, which have continued for over thirty years and have now evolved to collaborations on topics related to those covered in these lectures. Much of the clarity that this book may have is due in large part to many discussions, especially with Profs. Ernst and van Beijeren, and also with Drs Donald Jacobs and Harmen Bussemaker. It is a pleasure to thank Prof. Nico van Kampen for interesting discussions on quantum chaos and on linear response theory. I would also like to express my gratitude to Ms. Leonie Silkens for her help on many matters during my stay in Utrecht.

My colleagues at the University of Maryland, Mischa Brin and Garrett Stuck of the Department of Mathematics, and especially those in the chaos group, Edward Ott, Brian Hunt, Celso Grebogi, and James Yorke provided me with a first-rate education in dynamical systems theory and chaos, and have been encouraging of my efforts to relate their field to mine, statistical mechanics. My colleagues in the Institute for Physical Science and Technology, particularly Profs. Jan Sengers, Ted Kirkpatrick, Dave Thirumalai, and John Weeks, have been helpful indeed in welcoming me back to research after a period in administration. Thanks are also due to the Institute for Physical Science and Technology, the Department of Physics, the Office of the Dean of the College of Computer, Mathematical and Physical Sciences, and the Office of the Vice-President for Academic Affairs at the University of Maryland at College Park for various forms of financial and logistical support.

I especially want to thank Masao Yoshimura, Arnulf Latz, Rainer Klages, Charles Ferguson, Mihir Arjunwadkar and Kenneth Snyder for their considerable help in getting the book in its final form, for teaching me a great deal about LaTeX and computers in general, and for their valuable scientific support. Prof. John Weeks kindly provided Figure 2.1. I am indebted to Charles Ferguson for the figures of the cat map in Chapter 8. Thanks are also due to Prof. Michel Droz, Mr. Jerome Magnin, and the Department of Physics of the Univsersity of Geneva for their kind hospitality during May-June, 1998, when some parts of this book were written.

Much of what is new in this book is due to very happy scientific collaborations that I have had with Henk van Beijeren, Pierre Gaspard, Matthieu Ernst, E. G. D. Cohen, Shuichi Tasaki, Harald Posch, Rainer Klages, Arnulf Latz, Cecile Appert, Donald Jacobs, Christoph Dellago, Charles Ferguson, Debabrata Panja, and Thomas Gilbert. Prof. Predrag Cvitanović was kind enough to use a version of this text in a course at Northwestern University, and provided me with a number of corrections and helpful remarks. I thank him and his students at Northwestern for their suggestions and advice. Masao Yoshimura, Arnulf Latz, Rainer Klages, Ernest Barreto, Jane Gaily, Kenneth Snyder, Thomas Gilbert, Mihir Arjunwadkar, Karol Zyczkowski, Carl Dettmann, Juergen Vollmer, Támas Tél, Raul Rechtman, Pierre Gaspard, Luis Nasser, and David Urbach also read large parts of the text and made many valuable and significant suggestions for improvement. I am very indebted to Howard Weiss for a critical reading of the manuscript, which led to many improvements in the book, as well as for several important references to papers in the mathematical literature. Critical remarks by Jean Bricmont and anonymous referees led to a number of clarifications at various important points in the text and I thank Prof. Bricmont, especially, for his interesting and very stimulating thoughts on a number of matters.

There are two individuals to whom I owe a special debt of gratitude. Prof. Pierre Gaspard of the Université Libre de Bruxelles has aided me enormously in my understanding of the connections between nonequilibrium statistical mechanics and dynamical systems theory both through his writings and through our collaboration on a number of interesting topics. I wish to thank Prof. E. G. D. Cohen of the Rockefeller University who through many years of close and fruitful collaboration, has helped shape my understanding of irreversible processes.

The English translation of the Hebrew text from Genesis is used with permission of the Jewish Publication Society. The Hebrew phrase, 'tohu va'vohu', has entered a number of languages (English, French,...) as 'tohubohu', meaning 'chaos', 'disorder', or 'confusion'.

Finally, I would like to acknowledge support from the National Science Foundation under Grants No. PHY-93-21312 and PHY-96-00428.

A web site has been established for this book, at the time of publication. This web site can be accessed through the author's home page at http://www.ipst.umd.edu/dorfman.

J. R. Dorfman
College Park, Maryland
December, 1998

1
Nonequilibrium statistical mechanics

1.1 Introduction

Statistical mechanics is a very fruitful and successful combination of (i) the basic laws of microscopic dynamics for a system of particles with (ii) the laws of large numbers. This branch of theoretical physics attempts to describe the macroscopic properties of a large system of particles, such as one would find in a fluid or solid, in terms of the average properties of a large ensemble of mechanically identical systems which satisfy the same macroscopic constraints as the particular system of interest. The macroscopic phenomena that concern us in this book are those which fall under the general heading of irreversible thermodynamics, in general, or of fluid dynamics in particular. We shall be concerned with the second law of thermodynamics, more specifically, with the increase of entropy in irreversible processes. The fundamental problem is to reconcile the apparent irreversible behavior of macroscopic systems with the reversible, microscopic laws of mechanics which underly this macroscopic behavior. This problem has actively engaged physicists and mathematicians for well over a century.

1.2 The law of large numbers and the laws of mechanics

Many features of the solution to this problem were clear already to the founders of the subject, Maxwell, Boltzmann, and Gibbs, among others. The notion that equilibrium thermodynamics and fluid dynamics have a molecular basis is one of the central scientific advances of the 19th century. Of particular interest to us here is the work of Maxwell and Boltzmann, who tried to understand the laws of entropy increase in spontaneous natural processes on the basis of the classical dynamics of many-particle systems. Boltzmann's derivation, in 1872, of what is now known as the Boltzmann transport equation was a major step in the process of making the connection between molecular motions and irreversible thermodynamics. Boltzmann considered a dilute gas of particles interacting with short-range, central, pairwise forces, and obtained, using what appeared to be completely mechanical arguments, an equation for the distribution function, $F(\mathbf{r}, \mathbf{v}, t)$, of particles in a small region, $\delta\mathbf{r}$ about a point at position $\mathbf{r}$, with velocity in the range $\delta\mathbf{v}$ about velocity $\mathbf{v}$ at time t. This equation, which we will derive in the next chapter, has the interesting property that one can define a function of time, $H(t)$, in terms of the distribution function, which decreases monotonically in time, and reaches a constant value when the velocity distribution function is the Maxwell–Boltzmann equilibrium distribution system is spatially uniform. Furthermore, when evaluated for a system with this equilibrium distribution, the H-function is exactly $-S/k_{\mathrm{B}}$, where S is the thermodynamic entropy for an ideal gas, and k_{B} is Boltzmann's constant.

Consequently, Boltzmann had *almost* achieved the resolution of thermodynamics with mechanics, at least for dilute gases, by identifying this H-function with the negative of the thermodynamic entropy.

Objections were raised to Boltzmann's derivation of the laws of irreversible thermodynamics, based upon the time-reversal invariance of Newton's equations of motion and upon the Poincaré recurrence theorem. The former objection (called *Loschmidt's paradox*) says that if there is a motion of the gas that leads to a steady decrease of H with time, then there is certainly another allowed state of motion of the system, found by time reversal, in which H must increase. The second objection is a bit more subtle. Poincaré had proved that a bounded – in space and energy – mechanical system must typically have a recurrence property. That is, almost every (with the exception of a set of measure zero) initial state of an isolated, bounded, mechanical system will recur to within any specified accuracy, in the course of time. Of course, if H decreases over part of this motion, it must increase over some other part. This was referred to as *Zermelo's paradox*. The fact that this recurrence time may be much longer than the age of the universe is no escape from the argument that Boltzmann's derivation must contain some non-mechanical elements.

In fact, Boltzmann's derivation makes use of a stochastic argument called the *assumption of molecular chaos*, which allows an approximate calculation of the rates at which collisions are taking place in the gas. Nevertheless, as we discuss in later chapters, the derivation of the Boltzmann equation, the paradoxes surrounding it, and the modern ideas that have followed from it, are an essential part of nonequilibrium statistical mechanics, not only because of the deep and interesting conceptual problems involved, but also because the results of the Boltzmann equation are of great practical value in many areas of physics and engineering, and they need a firm foundation.

The recognition that the law of entropy increase must be something more than a consequence of Newton's laws led to the introduction of probabilistic ideas into this branch of physics. That is, the Boltzmann equation predicts things that are verified in laboratory experiments, despite the fact that this equation cannot be strictly correct; at least, not according to the laws of mechanics. One might say that the Boltzmann equation is 'probably' correct, rather than absolutely correct. It must describe a typical laboratory situation over times which are much longer than the time-scales of laboratory measurements. The H-theorem appears to hold for the typical behavior of a dilute gas, so the time-reversed motion of this gas, so important to Loschmidt, must correspond to a rare, very improbable, state of the gas. Moreover, the Poincaré recurrence time must be shown to be so large, compared to usual time-scales, that one will very likely never see such a recurrence. Since the reversal and recurrence objections are based entirely on mechanical principles, the introduction of probability arguments should be based upon the fact that macroscopic systems consist of large numbers of particles, and this fact should be coupled with a study of the dynamics of systems of large numbers of particles to provide a complete picture of irreversible processes.

The basic approach to the statistical mechanics of irreversible processes consists of three central themes:

1. One examines the average behavior of ensembles of mechanically identical systems. To do this, in classical mechanics at least, one constructs a phase-space, denoted Γ-space, with one coordinate axis for each canonical coordinate and one axis for each

canonical momentum, for the entire system. One point in Γ-space then represents the complete physical state of one system. One then describes the ensemble in terms of a phase-space distribution which gives the density of points in Γ-space. Then one calculates the average properties of this ensemble in terms of the phase-space distribution and its evolution in time.

2. One then needs to understand the mechanical behavior of a system of particles, that is, of each member of the ensemble of systems, well enough to identify those features of the dynamics which, under the proper circumstances, will allow one to demonstrate that the average macroscopic properties of the ensemble will satisfy the laws of irreversible thermodynamics, and approach their equilibrium values in the course of time.† This is what we will call the *mechanical problem of statistical mechanics.*

3. One needs to show that a laboratory system is *typical.* This means that the distribution of the values of macroscopic properties among the members of the ensemble is so sharply peaked at the average value that the average properties of the ensemble are overwhelmingly likely to be exhibited by any individual member of the ensemble. Of course, the solution of this problem depends on the laws of large numbers, and the fact that 'unlikely' behavior is characteristic of regions of phase-space with very small probability measure.

To illustrate this discussion, consider the behavior of a system of particles composed of the air molecules in a room, and take an initial state where all the molecules are in some corner of the room.‡ Now it is mechanically possible for the molecules to further compress themselves into the corner. However, it is much more likely, considering a reasonable probability measure on the phase-space, for a system like the one we are studying, for the molecules to start moving into the rest of the room. We also expect, from our experience with gases at room densities, that eventually the flow of molecules into the room will be described by the diffusion

† Of course, one has to specify what the 'proper circumstances' are.

‡ There is also a profound issue about the type of initial states we find or manufacture in laboratory systems and in the universe as a whole. This issue is essentially the question of the arrow of time, and we refer the reader to the references listed at the end of this chapter for a detailed discussion of this topic.

equation. In order to develop a theory to explain this behavior we have to consider both the mechanical and the statistical properties for systems of large numbers of particles. While our focus in this book will be on the mechanical properties, we need to keep the statistical properties in mind, since an understanding of the nonequilibrium behavior of large systems requires a careful analysis of both the statistical and the mechanical aspects of the system. This is also essential in order to make contact with laboratory experiments, which are always carried out on individual systems and not on large ensembles of similar systems.

It is a common practice now to study systems of a few particles using computer simulated molecular dynamics. By few particles, we mean numbers on the order of 10^2 to 10^4 particles, which must be compared to Avogadro's number, 10^{24}. Therefore it is also interesting to consider the statistical mechanics of small systems of particles, and to look at larger fluctuations about average values than the founders of statistical mechanics may have had in mind.

1.3 Boltzmann's ergodic hypothesis

Maxwell and Boltzmann were aware of the need to provide a better mechanical foundation for the second law of thermodynamics beyond the stochastic arguments advanced by Boltzmann. This gave rise to the notion of the *ergodic behavior* of a mechanical system which, from one point of view at least, is the starting point for all of equilibrium statistical mechanics. The main lines of the ergodic approach are based upon the following considerations: Boltzmann argued that, for systems with a large number of degrees of freedom, most of Γ-space† is taken up by regions where the macroscopic properties of systems in them have values very close to those we associate with thermodynamic equilibrium. Boltzmann then made the hypothesis that a mechanical system's trajectory in phase-space will spend *equal times* in regions of *equal phase-space measure.* If this is true, then any dynamical system will spend most of its time in phase-space regions where the values of the interesting macroscopic properties are extremely close to the equilibrium values. In other words, the macroscopic quan-

† 'Most', means in the sense of some natural probability measure.

tities are essentially constant on a very large part of phase-space. This would answer the mechanical paradoxes, since an approach to equilibrium would probably occur for the time-reversed motion of a mechanical system as well, and the Poincaré recurrence theorem would not be violated since, very occasionally, the systems phase point would come close to its initial value. Moreover, Boltzmann argued that if this hypothesis is correct, then the long-time average of a dynamical property of one system should be equal to the average value of that property taken with respect to a properly weighted ensemble density. Further, if the dynamical quantity is chosen properly, its ensemble average should correspond to the equilibrium thermodynamic value. For example, we can identify some mechanical property of the gas, the pressure, say, with the force per unit area exerted by the gas molecules on the walls of the container. Certainly a hypothetical microscopic measurement of this force per unit area would be a wildly fluctuating quantity. Due to the many orders of magnitude differences between the time-scales of microscopic dynamics and the macroscopic measurements, caused, in part, by the inertia of the laboratory devices, one is not far off in relating a macroscopic measurement to a time average of some microscopic property, over some macroscopic time interval. Next, one argues that if the mechanical motion of a large dynamical system is 'irregular enough', in a sense to be made more precise below, then, over a macroscopic time interval, the system will sample a very large number of the dynamical states available to it, consistent with the macroscopically imposed constraints of constant total energy and, perhaps, momentum. Under such circumstances, the macroscopic time average of some dynamical property can be equated to an ensemble average with a weight that is proportional to the fraction of the time that the system spends in each of the available dynamical states.

While Boltzmann's hypothesis, that the phase-space trajectory of a system spends an amount of time in a region that is proportional to its measure, is not at all forbidden by the laws of mechanics, it does place some heavy burdens on the dynamics. For instance, a dynamical system consisting of N particles, say, in d spatial dimensions, has $2dN$ constants of the motion, since the motion can be described by $2dN$ first-order differential equations, Hamilton's equations with solutions determined by $2dN$ initial constants. One combination of these constants is fixed by the total energy, and another by the initial point on the system's trajectory in phase-space. However, the phase points on the constant-energy surface differ in the particular values of almost all of the remaining constants of the motion. If the trajectory of one system is to visit an arbitrary region on the constant-energy surface, then the other, non-specified constants of the motion must be highly singular functions of the phase-space variables so that they take on almost all possible values in any arbitrarily small region about almost every point on the constant-energy surface.† Otherwise the system could not visit certain regions and the 'equal times in equal areas' rule would fail.

Granting for the moment the validity of the 'equal times in equal areas' rule, one can prove that the long-time average of some property of the system is then equal to an ensemble average, taken with respect to the so-called microcanonical distribution. From this, one can derive all of the results of statistical thermodynamics for classical systems. Of course, the problem of proving the ergodic character of a given mechanical system was left unsolved, as, for the most part, it still is today.

One simple objection to basing the foundation of equilibrium statistical mechanics on the presumed ergodic properties of mechanical systems is that laboratory experiments do not take an infinite amount of time, fortunately for experimentalists. Therefore one needs to show that over the amount of time of a laboratory experiment, the system's trajectory makes its way from a phase-space region where the macroscopic quantities do not have equilibrium values to a region where the values are those at equilibrium (essentially – there will always be some small fluctuations). The possibility of this happening, of course, depends on the dynamical properties of the system as well as on the sharpness of the phase-space distribution of the macroscopic quantities about their average values. The more irregular the motion, the more likely the system will be to wander about the phase-space and to sample a large enough region. Furthermore, it is impossible to isolate any mechanical system. Our laboratory system is con-

† In such cases one cannot think of such hypothetical functions as 'constants of the motion'. Instead one should restrict this term to analytic functions which are constant along a trajectory.

stantly subjected to outside perturbations. These perturbations may themselves make the system visit all regions of phase-space, even if the system, considered in isolation, is unable to do so. It is one of the tasks of modern statistical mechanics to try to sort out for a given system, how much of its apparent ergodic behavior can be attributed to its dynamical properties, how much is due to the inevitable external perturbations, and to address the question of whether global ergodic behavior is necessary at all for the foundations of statistical mechanics.

The main point to emphasize here is that for statistical mechanics one may only need a very restricted form of the ergodic hypothesis, where the 'equal times in regions of equal measure' rule holds in a space of low dimensions that characterizes only the dependences of the macroscopic variables on each other and on time, and then over time-scales on the order of laboratory times.† We shall emphasize this point at various places in our discussions. In fact, we shall see a few simple examples where projections of distribution functions onto low-dimensional phase-spaces approach equilibrium values on time-scales short compared to the time needed for full phase trajectories to sample the whole phase-space.

It is also worth mentioning that although our aim will be to characterize a kind of dynamical randomness, or chaotic behavior, of mechanical systems of interest to thermodynamics, randomness may have several sources, including chaotic dynamics, random external perturbations, and externally imposed randomness, such as is found in the random location of scatterers as electrons flow through an amorphous material. In any physical system, all of these sources of randomness must be identified and taken into account.

† This raises the question of the proper definition of a macroscopic variable. Usually these are the microscopic versions of the variables that appear in the thermodynamic or fluid dynamic equations. A discussion of a proper definition of these variables can be found in the book of Uhlenbeck and Ford mentioned at the end of the chapter.

1.4 Gibbs' mixing hypothesis

After the work of Maxwell and Boltzmann, Gibbs introduced the notion of a *mixing system.* Using the idea of mixing a drop of oil in an immiscible fluid, Gibbs argued that if one likened a small set of points in phase-space to the 'drop of oil', then the dynamical evolution of such a set would lead to its becoming uniformly mixed, *at least on a coarse-grained scale*, over the entire phase-space. Gibbs argued that the average behavior of such a small set would describe the typical behavior of a laboratory system. That is, since we are unable to totally specify the parameters needed to pick out just one trajectory in phase-space, we might reasonably try to characterize a typical, or generic, behavior of some appropriate set of trajectories. One might characterize the mixing property of a dynamical system by saying that infinitesimally close points in any small set (of positive probability measure) in phase-space will separate so rapidly that the set becomes 'strung out' over a large region of phase-space in a short amount of time compared to macroscopic times. If so, we will say that the system is mixing. In later chapters, we will characterize the rate of separation of nearby trajectories in phase-space by quantities called *Lyapunov exponents.* If a system possesses positive Lyapunov exponents, then there will be an exponential rate of separation of nearby trajectories, and we call such systems *chaotic.*†

We will give a precise definition of the mixing property in Chapter 5, and we will show that a mixing system is also ergodic. Moreover, as we usually describe the behavior of a set of trajectories in phase-space in terms of a phase-space distribution function, we will show that for a mixing system, this distribution function approaches, in a weak sense, the microcanonical equilibrium distribution function. Thus one can provide a mechanical foundation for both equilibrium and nonequilibrium statistical mechanics if one can prove that a given system is mixing. Of course, such proofs turn out to be nearly impossible for most systems of interest to physicists. Again, the mixing property in the full Γ-space may be much more than we need for describing the approach of a macro-

† It is worth mentioning that there are examples of mixing systems with no non-zero Lyapunov exponents. The concepts of ergodicity, mixing, and chaos can be quite subtle. See *Further reading* for references.

scopic system to equilibrium, and mixing in a space of lower dimensions may very well be enough for our purposes. We will see examples illustrating this point, and the time-scales involved, as we proceed. We will devote a large part of this book to describing some simple ergodic and mixing systems, generally systems of a low number of degrees of freedom, in order to illustrate the concepts, to show how they may be applied, with appropriate cautionary remarks, to deepen our understanding of the foundations of statistical mechanics and what might be required to make them secure.

1.5 Irregular dynamical motions

The irregular behavior of dynamical systems which might allow for the mixing or ergodic behavior we hope to find for statistical mechanics is not typical of the simple linear or integrable systems that we usually study in classical mechanics – such as harmonic oscillators or the two-body problem.† However, Poincaré, in his analysis of the gravitational three-body problems, showed that the motion of such systems can be very complicated indeed, and not easily described by ordinary mathematical notions of continuity, differentiability, and perturbation analysis. These systems characteristically exhibit the phenomenon of a *homoclinic tangle*, which is a complicated intersection of curves describing the system's motion in time, which remains complicated on an infinitesimally fine scale.‡ Such complicated behavior is certainly related to what one would expect for ergodic systems. It is worth mentioning here that Einstein based his objections to the Bohr–Sommerfeld rules for the quantization of classical systems upon the fact that these rules apply only to integrable mechanical systems.§ It is impossible to construct invariant tori for a complicated nonlinear mechanical system, like the classical helium atom. One might say

† The class of integrable systems includes, but is not restricted to systems in which the Hamilton–Jacobi equation can be solved by separation of variables.

‡ An excellent exposition of Poincaré's work can be found in the introduction by Goroff to the English translation of Poincaré's book *New Methods of Celestial Mechanics* (see Further reading).

§ The important point is that integrable systems can be analyzed in terms of invariant tori, to which the Bohr–Sommerfeld conditions can be applied.

that Einstein's observation was the beginning of modern quantum chaos theory.

1.6 Modern nonequilibrium statistical mechanics

Thus, in the early part of the 20th century there was a convergence of two lines of considerations for mechanical systems - one from statistical mechanics, and one from celestial mechanics - both of which suggested that most dynamical systems should have properties very different from linear ones treated in mechanics texts. Over the next decades much of the development of these ideas was carried on by mathematicians. The individual ergodic theorem of Birkhoff and the proof by Hopf of the ergodicity of trajectories on surfaces of constant negative curvature were reassuring to the physics community, but did not have much direct impact on new research directions on the foundations of statistical mechanics. Research in nonequilibrium statistical mechanics was largely devoted, then as now, to developing theories for the irreversible behavior of fluid and solid systems, using stochastic assumptions whenever necessary to obtain useful results. Theories of Brownian motion, sedimentation, and even stellar motions, typically use methods based upon clearly stochastic assumptions incorporated in the Langevin and Fokker–Planck equations. The nonequilibrium statistical mechanics of gases was extended to higher density by Bogoliubov, Green and Cohen, by using the Liouville equation to derive a set of equations (the BBGKY hierarchy equations†) for the reduced, or few-particle, distribution functions in a gas. Using equilibrium-like cluster expansion methods and assumptions about the smoothness and factorizability of the initial values of the reduced distribution functions, these authors were able to derive a formal generalization of the Boltzmann equation to higher densities than those considered by Boltzmann. We note in passing that virial expansions usually do not exist for quantities of importance in the nonequilibrium behavior, such as transport coefficients, of even moderately dense gases. This came as a surprise to workers in the field and is quite different from the situation in equilibrium statistical mechanics where virial expansions are quite useful.

† Born, Bogoliubov, Green, Kirkwood, and Yvon.

Another important advance in nonequilibrium statistical mechanics came with the development by Green and Kubo of the time–correlation-function method for expressing transport coefficients of fluid systems in terms of time integrals of equilibrium correlations of microscopic currents at two different times. The resulting expressions, known as Green–Kubo formulae, play a role similar to the partition function in equilibrium statistical mechanics. That is, the expressions are formal and to get useful results one must still resort to cluster expansions or some other approximate techniques.

Two new lines of development in the area of nonequilibrium statistical mechanics began in the 1940s. The first was an outgrowth of the introduction of computational methods into all areas of physics and engineering. In particular, Fermi, Ulam, and Pasta introduced the use of computers to study the possible ergodic behavior of nonlinearly coupled oscillator systems. To their surprise, no ergodic behavior was found, and only with the understanding of such systems, provided by the Kolmogorov–Arnold–Moser (KAM) theorem, was the situation of such oscillators clarified. Of course, the introduction of computer studies has had an importance that far transcends the applications to systems of oscillators. As a result of the studies by Fermi, Ulam, and Pasta, as well as by many other groups, a very deep analysis of mechanical systems began that was simply not possible before the advent of computers, both in scale and in the complicated behavior that could now be studied. Now the analyses of Poincaré of dynamical systems could be extended in many directions and the complicated behavior he discovered in celestial systems could be studied in great detail and for a wide variety of different systems. One of the consequences of such studies was a clarification of the notion of *chaos* in dynamical systems. In our context, *dynamical chaos* will be used to denote the sensitive dependence of the phase-space trajectories upon the initial conditions of the system. Many examples of the exponential separation of nearby trajectories in phase-space were discovered by computational studies, and the consequences of such sensitive behavior became the subject of chaos theory as a branch of the more general dynamical systems theory. Of course, one might speculate that Maxwell, Boltzmann, and Gibbs would have been greatly encouraged by chaos theory since, in their views, mechanical systems should typically show such irregular behavior, quite unlike that of simple harmonic oscillator systems. Major advances in the theory of dynamical systems were made by Smale, Bowen, and Ruelle, and by the group in Russia around Kolmogorov, and including Kolmogorov, Anosov, Arnold, Sinai, Pesin, Oseledets, and others. These authors developed the mathematical foundations for a rigorous treatment of the properties of what are now known as hyperbolic dynamical systems – which, very loosely speaking, are those which exhibit exponential separation of phase-space trajectories. Moreover, Sinai, Ruelle, and Bowen were able to show that there is a deep analogy between equilibrium statistical mechanics and methods used to characterize the mathematical properties of hyperbolic dynamical systems. The formal results suggested by this analogy, now called the *thermodynamic formalism*, have proved to be very powerful in the applications of dynamical systems theory to statistical mechanics, as might be expected, and will be used in this book as well.

A parallel but closely related development of interest to us was the argument of Krylov, long before the rise of chaos theory, that trajectories in phase-space for simple fluid systems would indeed separate exponentially with time. Krylov's arguments did not go beyond some simple, but convincing calculations, but they led eventually to the studies of Sinai and co-workers on the ergodic and mixing behavior of billiard systems. These authors were able to prove that some billiard systems are ergodic, mixing, and even stochastic-like in their behavior in time, despite their being time-reversible, deterministic dynamical systems. Moreover, such stochastic-like behavior could be analyzed in detail for simple systems with just a few degrees of freedom. These simple model systems provide a great deal of insight into how more complicated dynamical systems might actually behave, and how one could then provide a more substantial foundation for the validity of the Boltzmann equation and related stochastic-like equations commonly used to treat nonequilibrium systems. In this connection it is worth mentioning that Cohen and Gallavotti articulated this new dynamical approach to nonequilibrium statistical mechanics in what they called the 'chaotic hypothesis'. That is, one may treat fluid systems as if they were smooth, chaotic systems for the purpose of understanding their nonequilibrium properties, and for making general, but quantitative, statements about their behavior.† One important consequence of this hypothesis in the Gallavotti–Cohen fluctuation formula, described in Chapter 13.

1.7 Outline of this book

It is the purpose of this book to try to describe and discuss the developments given above and to provide the reader with some understanding of the current research topics and literature in this field. We will see that recent studies have led to a number of seemingly deep connections between the dynamical properties of a system, such as its Lyapunov exponents, and its transport properties. Recent studies of nonequilibrium stationary states, in particular, have shown that dynamical considerations are crucial for a detailed understanding of their properties. Here one needs to use ideas that have developed in the mathematical literature in connection with the properties of systems whose dynamics takes place on fractal structures of various sorts. Therefore, in order to understand recent work on nonequilibrium states, it is important to have a basic familiarity with standard methods, both in nonequilibrium statistical mechanics as well as in dynamical systems theory.

Chapters 2 through 6 of the book are designed to provide the background necessary for an understanding of by now standard methods in nonequilibrium statistical mechanics. They begin with a traditional discussion of the Boltzmann transport equation, which still today is the centerpiece of our understanding of the hydrodynamic properties of fluids. After the Boltzmann equation and the assumptions made in its derivation are discussed, more general approaches to transport theory such as the Green–Kubo time-correlation-function method are considered. Armed then with the basic tools of nonequilibrium statistical mechanics, the student is introduced to the fundamentals of chaos theory. The baker's transformation and, later the Arnold cat map are used as basic paradigms for understanding many aspects of the foundations of transport theory. These transformations are simple enough that they can be readily understood, but complex enough that many

† By 'smooth, chaotic' systems, we will mean Anosov, hyperbolic systems to be described in Chapter 9.

features of the theory of irreversible processes can be illustrated with their aid. The baker's transformation and the Arnold cat map take place on a phase-space of only two dimensions, yet they are instructive models for what might happen in a much larger phase-space, if one were to consider a projection of trajectories in the full phase-space for a large system onto a subspace of lower dimension. For these models one can easily see the approach to equilibrium of projected distribution functions on time-scales shorter than those needed for establishing the ergodic or mixing behavior of the system in the full phase-space.

Chapters 7 through 18 of the book discuss recent developments in this area and are intended to form an introduction to the current research literature on transport theory in general, and on nonequilibrium steady states, in particular. We present two approaches to transport theory which have developed over the past few years, and which are based on the notions of chaos theory:

1. The escape-rate or chaotic scattering theory formalism for computing transport coefficients developed by Gaspard, Nicolis, and co-workers, and,
2. The Gaussian thermostat method for computing transport coefficients, developed by Nosé, Hoover, Evans, Morriss, Posch, Cohen, and co-workers.

These methods are of crucial interest to us here since they allow one to directly relate macroscopic transport quantities, such as transport coefficients, to microscopic dynamical quantities like Lyapunov exponents and Kolmogorov–Sinai entropies. These two topics, as well as the fluctuation theorem of Gallavotti and Cohen discussed in Chapter 13, are, in my opinion, the first glimmers of what can be expected to be a new line of research in transport theory relating macroscopic properties of large systems to the properties of the underlying microscopic dynamics. The mathematical description of the phase-space structures that are important for transport theory in the approaches considered here requires the use of what have come to be known as SRB (Sinai–Ruelle–Bowen) measures and as Gibbs measures. The SRB measures appear naturally in the description of thermostatted systems, and are smooth measures in some directions of the phase-space and may be irregular or singular in other directions. To describe the mathematical properties of the fractal structures called repellers that appear in the escape-rate formalism, we need a more general class of measures, called Gibbs measures. Gibbs measures include SRB measures as a subclass, but are general enough to describe the properties of sets which are fractal in all directions. Chapter 13 is devoted to a discussion of these measures. Then we discuss a simple example of fractal forms that appear in the Green–Kubo formulation of transport phenomena. These forms are amusing and interesting in themselves, but should also be seen as a precursor to a more sophisticated approach to transport theory involving so-called Ruelle–Pollicott resonances. The latter are not discussed in this book, but we hope that the reader will not be surprised by what she or he finds in the literature, after reading this chapter. We then explore one further method for analyzing nonequilibrium states, namely the periodic orbit expansion method. This powerful method is used quite often in the literature to evaluate transport coefficients and related quantities and forms the basis for one approach to quantum chaos theory. A student entering this field ought to be familiar with the periodic orbit method. Chapter 17 is devoted to assembling the various mathematical results and physical arguments of the earlier chapters into a coherent picture of our understanding of the emergence of irreversible behavior in fluid systems. Here we again illustrate the point that projected, or reduced, distribution functions can approach equilibrium values on a much shorter time-scale than that needed for ergodicity or mixing of the system in phase-space. In Chapter 18, the Boltzmann equation returns in a new guise – as a basic method for computing the quantities, such as Lyapunov exponents, that arise naturally in a dynamical systems approach to random systems.

Lastly, Chapter 19 concludes with a brief guide to the very recent literature. Each chapter concludes with a guide to the literature and sources of the material presented in the chapter, and often a few exercises designed to stimulate the reader.

I have not made any attempts at mathematical rigor, and I occasionally use terms in a way that differs from the usual usage in the mathematical literature. In particular, I use the term *smooth* to describe a function which is continuous, and has enough derivatives so that what I say about it makes sense. In the mathematical literature, smooth describes a function that is differentiable an arbitrarily large number of times. Here, an *integrable function* will always mean a Lebesgue integrable function. Additionally, I often use the term *hyperbolic system* to mean what is usually termed an *Anosov system.* The differences between *hyperbolic*, *Anosov*, and *Axiom-A* systems are briefly outlined in Chapter 9. This is done primarily to assist the readers as they venture into the more technical literature, where these differences can often be crucial parts of an argument.

It is my hope that this book will serve to introduce students to this new area of research, to show that there are many deep, unsolved problems waiting for clever people to formulate them precisely and to solve them, and finally, perhaps, to show that the basic ideas provided by the founders of this subject – Maxwell, Boltzmann, and Gibbs – are beginning to be understood and verified by the mathematical and physical researches of the present day – one hundred years after these ideas were first developed.

In preparing first the lecture notes, and more recently this text, I relied upon a number of basic reference works. Most of them are listed in the bibliography, but there are a few sources that I found exceptionally helpful either as introductions to dynamical systems and chaos theory or to its applications to statistical mechanics. These I list here to acknowledge their importance for my own understanding of the subject as well as to call attention to their special relevance to the subjects covered in this book. They are:

1. E. Ott, *Chaos in Dynamical Systems*, Cambridge University Press, Cambridge (1992).
2. *Microscopic Simulations of Complex Hydrodynamical Phenomena*, edited by M. Maréschal and B. Holian, Plenum Press, New York (1992).
3. D. Ruelle and J-P. Eckmann, Ergodic theory of chaos and strange attractors, *Reviews of Modern Physics*, **57**, 617 (1985).
4. J. L. Lebowitz and O. Penrose, Modern ergodic theory, *Physics Today*, February 1973, pp. 23–29.
5. A. Lasota and M. C. Mackey, *Chaos, Fractals, and Noise*, 2nd edition., Springer-Verlag, New York (1994).
6. A. Katok and B. Hasselblatt, *Introduction to the Modern Theory of Dynamical Systems*, Cambridge University Press, Cambridge (1995).

7. P. Gaspard, *Chaos, Scattering, and Statistical Mechanics,* Cambridge University Press, Cambridge (1998).

A number of important topics in modern nonequilibrium statistical mechanics are not covered here. Such topics would include pattern formation, spatiotemporal chaos, driven diffusive systems, granular flows, theories of glass formation, and the mathematical physics of hydrodynamic processes of large systems. It is clearly not possible to cover so many topics in one introductory book, so I hope that the interested reader will consult the references listed below and those at the end of other chapters.

1.8 Further reading

A good discussion of the picture developed by Boltzmann, Maxwell, and Gibbs for the foundations of statistical mechanics can be found in the book by Uhlenbeck and Ford [UF63], among others. More recent surveys of Boltzmann's thoughts with an eye toward modern developments are given in papers by Cohen [Coh97b], by Gallavotti [Gal95], by Lebowitz [Leb93], and by Bricmont [Bri95]. Lebowitz and Bricmont both give careful discussions of the role of the statistical element in statistical mechanics, and Gallavotti provides an interesting discussion of the meaning of Boltzmann's ergodic hypothesis. A clear discussion of a wide range of topics, including statistical mechanics, ergodic theory, and chaotic dynamics can be found in a small book by Ruelle [Rue91]. Hunt and Yorke [HY93] have looked at Maxwell's writings and give a very good description of his very modern ideas, and those of Poincaré, on the chaotic foundations of statistical mechanics.

The role of initial low-entropy states, especially at the time of the 'big bang', for the apparent irreversibility of the universe is discussed by Feynman [Fey67] and by Penrose [Pen89], as well as by Lebowitz [Leb93].

Some of the recent literature on nonequilibrium statistical mechanics has been incorporated into the modern textbook literature. The books by Spohn [Spo91], by DeMasi and Presutti [DP91], by McLennan [McL89], and by Balescu [Bal97] are devoted to this subject. In particular, they discuss the derivation of hydrodynamic equations from microscopic theories, and discuss the role of the thermodynamic limit for obtaining these equations. The book by Spohn, among many other things, has a very clear discussion of microscopic reversibility and the Boltzmann transport equation. The two-volume set by Toda *et al.* [TKS92, KTH91] has a fairly extensive discussion of nonequilibrium statistical mechanics, as do the texts by Ma [Ma85] and by Reichl [Rei98]. The 19th century history of this subject is covered in detail in the book by Brush [Bru76], and the more recent developments in kinetic theory are discussed in his three-volume series on kinetic theory [Bru72]. Of course Boltzmann's book, in its translation by Brush is essential reading [Bol95]. A classic discussion of the assumptions used in the early development of nonequilibrium statistical mechanics by Boltzmann is the Encyclopedia article of P. and T. Ehrenfest [PE59].

The development of a systematic kinetic theory for dense gases, due to Bogoliubov, Green, and Cohen is summarized in the article by Dorfman and van Beijeren [DvB77], with a large bibliography included.

Surveys of many recent developments in nonequilibrium statistical mechanics are to be found in the books of Spohn and of DeMasi and Presutti mentioned above, as well as in papers collected by Berne [Ber77], in the series *Fundamental Problems in Statistical Mechanics* [CvB*et al.*nt], and in the series *Studies in Statistical Mechanics* [UdBL$^+$nt]. The book of Boon and Yip [BY91] describes the application of kinetic and hydrodynamic methods to a study of molecular processes. In addition, there are a number of survey articles of very recent developments in the papers by Ernst [Ern98], Cohen [Coh93], Dorfman [Dor81], Dorfman, Kirkpatrick, and Sengers [DKS94], Pomeau and Resibois [PR75], Halperin and Hohenberg [HH77], and Cross and Hohenberg [CH93], among many others.

The classic book of Poincaré on celestial mechanics exists in an English translation [Poi93], and, as mentioned earlier, the introduction by Goroff is quite valuable. An interesting discussion of Einstein's prescient views on the Bohr–Sommerfeld quantization rules can be found in the book by Gutzwiller [Gut90]. Much of the work of the Russian school on dynamical systems theory is collected in the volumes edited by Sinai [Sin89, Sin91], which include English translations of the fundamental papers of Sinai,

Oseledets, Pesin, Chernov, and Bunimovich. Some other essential papers are those of Anosov [Ano67], Anosov and Sinai [AS67], Katok [Kat81], and Wojtkowski [Woj85].

A survey of more recent work on Boltzmann's ergodic hypothesis can be found in the paper of Szász [Szá93], and a more mathematical discussion of the ergodic properties of hard-sphere-type systems in the paper by Liverani and Wojtkowski [LW95]. Examples of mixing systems with zero Lyapunov exponents can be found in the papers of Marcus [Mar77, MB77, Mar78] and in the paper by Pesin [Pes77]. The close connection between equilibrium statistical mechanics and the formal structure of dynamical systems theory, the so-called *thermodynamic formalism* is discussed by Sinai [Sin72], Bowen [Bow75, BR75, Bow78b], Ruelle [Rue78], and in the recent text of Beck and Schlögl [BS93]. We will use this formalism extensively in the later chapters of this book. A nice introduction to the mathematical foundations of dynamical systems theory is given by Lanford [Lan83], and a careful and clear discussion of the dynamical issues arising in the theory of nonequilibrium stationary states has been given in a recent set of notes by Ruelle [Rue98]. The Fermi–Pasta–Ulam model and the issues of its ergodicity, or lack thereof, are discussed in the book of Jackson [Jac90] and in the collection of papers edited by MacKay and Meiss [MM87]. The paper of Berry [Ber87] is especially illuminating. The chaotic hypothesis and the Gallavotti–Cohen fluctuation formula are discussed in two papers, [GC95a,GC95b].

2
The Boltzmann equation

2.1 Heuristic derivation

In 1872, Boltzmann introduced the basic equation of transport theory for dilute gases. His equation determines the time-dependent position and velocity distribution function for the molecules in a dilute gas. We begin by presenting Boltzmann's derivation of this equation in order to illustrate both the equation as well as some of the problems that arise in the course of its derivation.

Consider a large vessel of volume V, containing N molecules which interact with central, pairwise additive, repulsive forces. † The latter requirement allows us to avoid the complications of long-lived 'bound' states of two molecules which, though interesting, are not central to our discussion here. We suppose that the pair potential has a strong repulsive core and a finite range a. A typical pair potential is illustrated in Fig. 2.1. The gas is taken to be dilute, meaning that the volume per particle is much greater than the size of a particle, $V/N \gg a^3$, or, equivalently, $na^3 \ll 1$, where $n = N/V$ is the number density of the particles.

Now we define a distribution function, $F(\mathbf{r}, \mathbf{v}, t)$, for the gas over a 6-dimensional position and velocity space, $(\mathbf{r}, \mathbf{v})$, such that

$$F(\mathbf{r}, \mathbf{v}, t)\delta\mathbf{r}\delta\mathbf{v} \equiv \text{the number of particles in } \delta\mathbf{r}\delta\mathbf{v} \text{ around } \mathbf{r} \text{ and } \mathbf{v} \text{ at time } t. \tag{2.1}$$

$F(\mathbf{r}, \mathbf{v}, t)$ is the particle density in a 6-dimensional, one-particle phase-space, often referred to as μ-space. To get an equation for $F(\mathbf{r}, \mathbf{v}, t)$, we take a region $\delta\mathbf{r}\delta\mathbf{v}$ about a point $(\mathbf{r}, \mathbf{v})$ in this space, that is large enough to contain a lot of particles, but small compared to the range of variation of F.

There are four mechanisms that change the number of particles in this region. The particles can:

† We will use the terms *molecule* and *particle* interchangeably throughout the book.

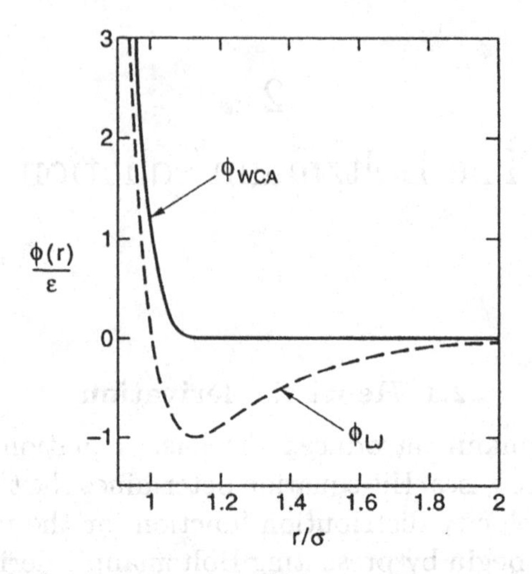

Fig. 2.1 Typical pair potentials. Illustrated here are the Lennard-Jones pair potential, ϕ_{LJ}, and the associated Weeks–Chandler–Anderson potential, ϕ_{WCA}, which gives the same repulsive force as the Lennard–Jones potential. The relative separation coordinate is scaled by the distance σ, the point at which ϕ_{LJ} first passes through zero, and the energy axis is scaled by the well depth, ε.

1. flow into or out of $\delta\mathbf{r}$, the *free-streaming term*,
2. leave the $\delta\mathbf{v}$ region as a result of a direct collision, the *loss term*,
3. enter the $\delta\mathbf{v}$ region after a restituting collision, the *gain term*, and
4. collide with the wall of the container (if the region contains part of the walls), the *wall term*.

Assume there is a time-scale δt which is sufficiently long that a great number of collisions can take place in each cell $\delta\mathbf{r}\,\delta\mathbf{v}$ in time δt, but which is too short for particles to cross a cell of size $\delta\mathbf{r}$. Then the change in the number of particles in $\delta\mathbf{r}\,\delta\mathbf{v}$ in time δt can be written as

$$[F(\mathbf{r},\mathbf{v},t+\delta t)-F(\mathbf{r},\mathbf{v},t)]\,\delta\mathbf{r}\delta\mathbf{v}=\Gamma_{\mathrm{f}}-\Gamma_{-}+\Gamma_{+}+\Gamma_{\mathrm{w}} \tag{2.2}$$

where Γ_{f}, Γ_{-}, Γ_{+}, and Γ_{w} represent the changes in F due to the four mechanisms listed above, respectively. We suppose that each particle in the small region suffers at most one collision during the time interval δt, and calculate the change in F.

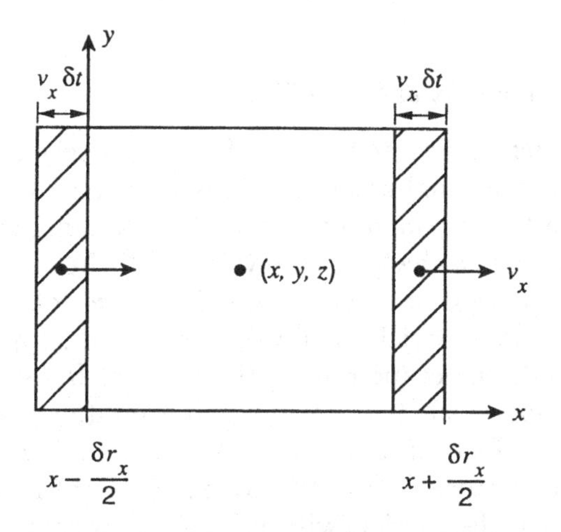

Fig. 2.2 A schematic illustration of flow into and out of a small region. The hatched areas represent regions where particles enter and leave $\delta\mathbf{r}\delta\mathbf{v}$ in time δt.

The computation of Γ_f is relatively straightforward. We simply consider the free flow of particles into and out of the region in time δt. An expression for this flow in the x-direction, for example, can be obtained by considering two thin layers of size $v_x\,\delta t\,\delta r_y\,\delta r_z$ that contain particles that move into or out of a cell with its center at (x, y, z) in time δt (see Fig. 2.2).

The free streaming term can be written as the difference between the number of particles entering and leaving the small region in time δt. Consider, for example, a cubic cell and look at the faces perpendicular to the x-axis. The flow of particles across the faces at $x - \frac{1}{2}\delta r_x$ and at $x + \frac{1}{2}\delta r_x$ is

$$\begin{aligned}\Gamma_\mathrm{f}^{(x)} &= v_x\,\delta t\,\delta r_y\,\delta r_z\,\delta\mathbf{v} \times \\ &\left[F\left(x - \frac{1}{2}\delta r_x, y, z, \mathbf{v}, t\right) - F\left(x + \frac{1}{2}\delta r_x, y, z, \mathbf{v}, t\right)\right],\end{aligned} \tag{2.3}$$

and similar expressions exist for the y- and z-directions. The function F is supposed to be sufficiently smooth that it can be expanded in a Taylor series around (x, y, z). The zeroth-order terms between the parentheses cancel and the first-order terms add up. Neglecting terms of order δ^2 and higher and summing over all directions then yields

$$\Gamma_{\mathrm{f}} = -\delta\mathbf{v}\,\delta t\,\delta\mathbf{r}\,(\mathbf{v}\cdot\nabla)\,F(\mathbf{r},\mathbf{v},t). \tag{2.4}$$

Next we consider the computation of the loss term, Γ_-. We need to calculate the number of collisions suffered by particles with velocity $\mathbf{v}$ in the region $\delta\mathbf{r}\,\delta\mathbf{v}$ in time δt, assuming that each such collision results in a change of the velocity of the particle. We carry out the calculation in several steps. First, we focus our attention on a particular particle with velocity $\mathbf{v}$, and suppose that it is going to collide sometime during the interval $[t, t+\delta t]$ with a particle with velocity $\mathbf{v}_1$. Now set up a coordinate system with origin at the center of the particle with velocity $\mathbf{v}$, and with z-axis directed along the vector $\mathbf{g} = \mathbf{v}_1 - \mathbf{v}$. By examining Fig. 2.3, one can easily see that if the particle with velocity $\mathbf{v}_1$ is somewhere at time t within the *collision cylinder* illustrated there, with volume $|\mathbf{v}_1 - \mathbf{v}|\,\pi a^2\,\delta t$, this particle will collide sometime during the interval $[t, t+\delta t]$ with the particle with velocity $\mathbf{v}$, if no other particles interfere, which we assume to be the case. These collision cylinders will be referred to as $(\mathbf{v}_1, \mathbf{v})$-collision cylinders. We also ignore the possibility that the particle with velocity $\mathbf{v}_1$ might, at time t, be somewhere within the action sphere of radius a about the center of the velocity-$\mathbf{v}$ particle, since such events lead to terms that are of higher order in the density than those we are considering here, and such terms do not even exist if the duration of a binary collision is strictly zero, as would be the case for hard spheres, for example.

We now compute Γ_- by noting the following:

- The number of $(\mathbf{v}_1, \mathbf{v})$-collision cylinders in the region $\delta\mathbf{r}\,\delta\mathbf{v}$ is equal to the number of particles with velocity $\mathbf{v}$ in this region, $F(\mathbf{r},\mathbf{v},t)\,\delta\mathbf{r}\,\delta\mathbf{v}$.
- Each $(\mathbf{v}_1, \mathbf{v})$-collision cylinder has the volume given above, and the total volume of these cylinders is equal to the product of the volume of each such cylinder with the number of these cylinders, that is $F(\mathbf{r},\mathbf{v},t)\,|\mathbf{v}_1 - \mathbf{v}|\,\pi a^2\,\delta\mathbf{r}\,\delta\mathbf{v}\,\delta t$.
- If we wish to know the number of $(\mathbf{v}_1, \mathbf{v})$-collisions that actually take place in this small time interval, we need to know exactly where each particle is located and then follow the motion of *all* the particles from time t to time $t+\delta t$. In fact, this is what is done in

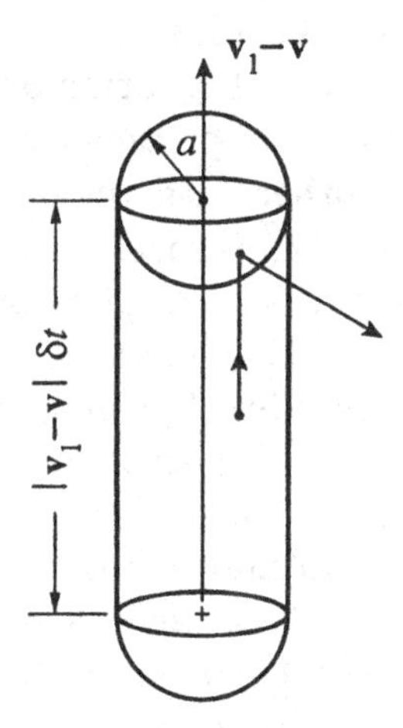

Fig. 2.3 The $(\mathbf{v}_1, \mathbf{v})$-collision cylinder. The sphere has a radius a which is the range of the forces. For hard sphere molecules, a is the diameter of the molecules.

computer simulated molecular dynamics. We wish to avoid this exact specification of the particle trajectories, and instead carry out a plausible argument for the computation of Γ_-. To do this, Boltzmann made the following assumption, called the *Stosszahlansatz*:

(Stosszahlansatz) *The total number of $(\mathbf{v}_1, \mathbf{v})$-collisions taking place in δt equals the total volume of the $(\mathbf{v}_1, \mathbf{v})$-collision cylinders times the number of particles with velocity $\mathbf{v}_1$ per unit volume.*

After integration over $\mathbf{v}_1$, we obtain

$$\Gamma_- = \delta\mathbf{r}\,\delta\mathbf{v}\,F(\mathbf{r},\mathbf{v},t)\int d\mathbf{v}_1\,\delta t\,\pi a^2\,|\mathbf{v}_1-\mathbf{v}|\,F(\mathbf{r},\mathbf{v}_1,t). \qquad (2.5)$$

The gas has to be dilute because the collision cylinders are assumed not to overlap, and also because collisions between more than two particles are neglected. Also it is assumed that F hardly changes over $\delta\mathbf{r}$ so that the distribution functions for both colliding particles can be taken at the same position $\mathbf{r}$.

The assumptions that go into the calculation of Γ_- are referred to collectively as the *assumption of molecular chaos.* In this context, this assumption says that the probability that a pair of particles with given velocities will collide can be calculated by considering each particle separately and ignoring any correlation between the probability for finding one particle with velocity $\mathbf{v}$ and the probability for finding another with velocity $\mathbf{v}_1$ in the region $\delta\mathbf{r}$.

For the construction of Γ_+, we need to know how two particles can collide in such a way that one of them has velocity $\mathbf{v}$ after the collision. The answer to this question can be found by a more careful examination of the 'direct' collisions which we have just discussed. To proceed with this examination, we note that the factor πa^2 appearing in (2.5) can also be written as an integral over the impact parameters and azimuthal angles of the $(\mathbf{v}_1, \mathbf{v})$ collisions. That is, $\pi a^2 = \int b\,db \int d\epsilon$, where b, the impact parameter, is the initial distance between the center of the incoming $\mathbf{v}_1$-particle and the axis of the collision cylinder (z-axis), and ϵ is the angle between the x-axis and the position of particle 2 in the x–y plane. Here $0 \leq b \leq a$, and $0 \leq \epsilon \leq 2\pi$. The laws of conservation of linear momentum, angular momentum, and energy require that both the impact parameter b, and $|\mathbf{g}| = |\mathbf{v}_1 - \mathbf{v}|$, the magnitude of the relative velocity, be the same before and after the collision. To see what this means let's follow the two particles through and beyond a direct collision. We denote all quantities after the collision by primes. The conservation of momentum

$$\mathbf{v}_1 + \mathbf{v} = \mathbf{v}_1' + \mathbf{v}'$$

implies, after squaring and using conservation of energy,

$$v_1^2 + v^2 = v_1'^2 + v'^2,$$

that

$$\mathbf{v}_1 \cdot \mathbf{v} = \mathbf{v}_1' \cdot \mathbf{v}'.$$

By multiplying this result by a factor of -2, and adding the result to the conservation of energy equation, one easily finds $g = g' = |\mathbf{v}_1' - \mathbf{v}'|$. This result, taken together with conservation of angular momentum, $\mu g b = \mu g' b'$, where $\mu = \frac{1}{2}m$ is the reduced mass of the two-particle system, shows that b is also conserved, $b = b'$. This is illustrated in Fig. 2.4.

Next, we denote the line between the centers of the two particles at the point of closest approach by the unit vector $\hat{\mathbf{k}}$. In Fig. 2.4, it can also be seen that the vectors $-\mathbf{g}$ and $\mathbf{g}'$ are each other's mirror images in the direction of $\hat{\mathbf{k}}$ in the plane of the trajectory

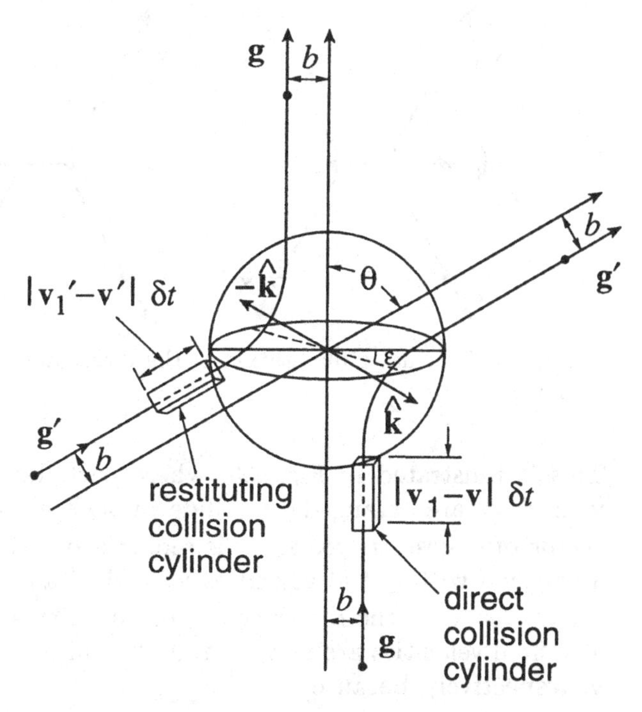

Fig. 2.4 Direct and restituting collisions in the relative coordinate frame. The collision cylinders, as well as the appropriate scattering and azimuthal angles, are illustrated.

of particles:

$$\mathbf{g}' = \mathbf{g} - 2\left(\mathbf{g} \cdot \hat{\mathbf{k}}\right)\hat{\mathbf{k}}, \tag{2.6}$$

and thus $(\mathbf{g} \cdot \hat{\mathbf{k}}) = -(\mathbf{g}' \cdot \hat{\mathbf{k}})$. Together with conservation of momentum this gives

$$\begin{aligned} \mathbf{v}_1' &= \mathbf{v}_1 - \left(\mathbf{g} \cdot \hat{\mathbf{k}}\right)\hat{\mathbf{k}} \\ \mathbf{v}' &= \mathbf{v} + \left(\mathbf{g} \cdot \hat{\mathbf{k}}\right)\hat{\mathbf{k}}. \end{aligned} \tag{2.7}$$

The main point of this argument is to show that if particles with velocities $\mathbf{v}'$ and $\mathbf{v}_1'$ collide in the right geometric configuration with impact parameter b, such a collision will result in one of the particles having the velocity of interest, $\mathbf{v}$, after the collision. These kinds of collisions which produce particles with velocity $\mathbf{v}$, contribute to Γ_+, and are referred to as 'restituting' collisions.

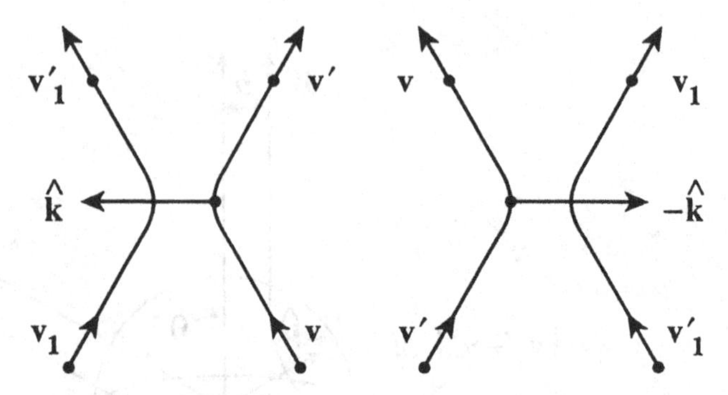

Fig. 2.5 Schematic illustration of the direct and restituting collisions.

This is illustrated in Fig. 2.5, where particles having velocities $\mathbf{v}'$ and $\mathbf{v}_1'$ are arranged to collide in such a way that the unit vector of closest approach, $\hat{\mathbf{k}}$, is replaced by $-\hat{\mathbf{k}}$. Consider, then, a collision with initial velocities $\mathbf{v}_1'$ and $\mathbf{v}'$ and the same impact parameter as in the direct collision, but with $\hat{\mathbf{k}}$ replaced by $-\hat{\mathbf{k}}$. The final velocities are now $\mathbf{v}_1''$ and $\mathbf{v}''$, which are equal to $\mathbf{v}_1$ and $\mathbf{v}$, respectively, because

$$\mathbf{v}_1'' = \mathbf{v}_1' - \left(\mathbf{g}' \cdot \hat{\mathbf{k}}\right) \hat{\mathbf{k}} = \mathbf{v}_1' + \left(\mathbf{g} \cdot \hat{\mathbf{k}}\right) \hat{\mathbf{k}} = \mathbf{v}_1, \tag{2.8}$$

and

$$\mathbf{v}'' = \mathbf{v}' - \left(\mathbf{g} \cdot \hat{\mathbf{k}}\right) \hat{\mathbf{k}} = \mathbf{v}. \tag{2.9}$$

Thus the *increase* of particles in our region due to restituting collisions with an impact parameter between b and $b + db$ and azimuthal angle between ϵ and $\epsilon + d\epsilon$ (see Fig. 2.4) can be obtained by adjusting the expression for the *decrease* of particles due to a 'small' collision cylinder:

$$\begin{aligned} &\text{Loss:} \quad \delta t\, b\, db\, d\epsilon\ |\mathbf{g}|\ F(\mathbf{r}, \mathbf{v}, t)\, F(\mathbf{r}, \mathbf{v}_1, t)\, \delta\mathbf{v}\, \delta\mathbf{v}_1\, \delta\mathbf{r} \\ &\text{Gain:} \quad \delta t\, b\, db\, d\epsilon\ |\mathbf{g}'|\ F(\mathbf{r}, \mathbf{v}', t)\, F(\mathbf{r}, \mathbf{v}_1', t)\, \delta\mathbf{v}_1'\, \delta\mathbf{v}'\, \delta\mathbf{r}, \end{aligned}$$

where b has to be integrated from 0 to a, and ϵ from 0 to 2π. Also, by considering the Jacobian for the transformation to relative and center of mass velocities, one easily finds that $d\mathbf{v}_1\, d\mathbf{v} = d\mathbf{V}\, d\mathbf{g}$, where $\mathbf{V}$ is the velocity of the center of mass of the two colliding particles with respect to the container. After a collision, $\mathbf{g}$ is rotated in the center of mass frame, so the Jacobian of the transformation $(\mathbf{V}, \mathbf{g}) \to (\mathbf{V}', \mathbf{g}')$ is unity and $d\mathbf{V}\, d\mathbf{g} = d\mathbf{V}'\, d\mathbf{g}'$. So

$$d\mathbf{v}_1\, d\mathbf{v} = d\mathbf{V}\, d\mathbf{g} = d\mathbf{V}'\, d\mathbf{g}' = d\mathbf{v}_1'\, d\mathbf{v}'. \qquad (2.10)$$

Now we are in the correct position to compute Γ_+, using exactly the same kinds of arguments as in the computation of Γ_-, namely, the construction of collision cylinders, computing the total volume of the relevant cylinders and again making the *Stosszahlansatz*. Thus, we find that

$$\Gamma_+ = \iiint d\mathbf{v}_1'\, b\, db\, d\epsilon\, |\mathbf{v}_1' - \mathbf{v}'|\, F(\mathbf{r}, \mathbf{v}', t)\, F(\mathbf{r}, \mathbf{v}_1', t)\; \delta\mathbf{r}\, \delta\mathbf{v}'\, \delta t. \qquad (2.11)$$

For every value of the velocity $\mathbf{v}$, the velocity ranges $d\mathbf{v}_1'\, \delta\mathbf{v}'$ in the above expression are only over that range of velocities $\mathbf{v}'$, $\mathbf{v}_1'$ such that particles with velocity in the range $\delta\mathbf{v}$ about $\mathbf{v}$ are produced in the $(\mathbf{v}', \mathbf{v}_1')$-collisions. If we now use the equalities (2.10) as well as the fact that $|\mathbf{g}| = |\mathbf{g}'|$, we can write

$$\Gamma_+ = \iiint d\mathbf{v}_1\, b\, db\, d\epsilon\, |\mathbf{v}_1 - \mathbf{v}|\; F(\mathbf{r}, \mathbf{v}', t) F(\mathbf{r}, \mathbf{v}_1', t)\, \delta\mathbf{r}\, \delta\mathbf{v}\, \delta t. \qquad (2.12)$$

The term describing the interaction with the walls, Γ_{w}, is discussed in the paper by Dorfman and van Beijeren mentioned at the end of this chapter. Here we will not go into the detailed properties of this term, but will quote results as needed.

Finally, all of the Γ-terms can be inserted in (2.2), and dividing by $\delta t\, \delta\mathbf{r}\, \delta\mathbf{v}$ gives the Boltzmann transport equation:

$$\frac{\partial F(\mathbf{r}, \mathbf{v}, t)}{\partial t} + \mathbf{v} \cdot \nabla F(\mathbf{r}, \mathbf{v}, t) = J(F, F) + T_{\mathrm{w}}, \qquad (2.13)$$

where $J(F, F) = \iiint d\mathbf{v}_1 b\, db\, d\epsilon |\mathbf{v}_1 - \mathbf{v}|\, [F' F_1' - F_1 F]$.

The primes and subscripts on the Fs refer to their velocity arguments, and the primed velocities in the gain term should be regarded as functions of the unprimed quantities according to (2.7). The term describing the effect on F of the collisions of the particles with the walls, T_{w}, is given in detail in the paper of Dorfman and van Beijeren. It is often convenient to rewrite the integral over the impact parameter and the azimuthal angle as an integral over the unit vector $\hat{\mathbf{k}}$ as

$$g b\, db\, d\epsilon = B(\mathbf{g}, \hat{\mathbf{k}})\, d\hat{\mathbf{k}}, \qquad (2.14)$$

where

$$d\hat{\mathbf{k}} = \sin(\pi - \psi)\, d(\pi - \psi)\, d\epsilon, \qquad (2.15)$$

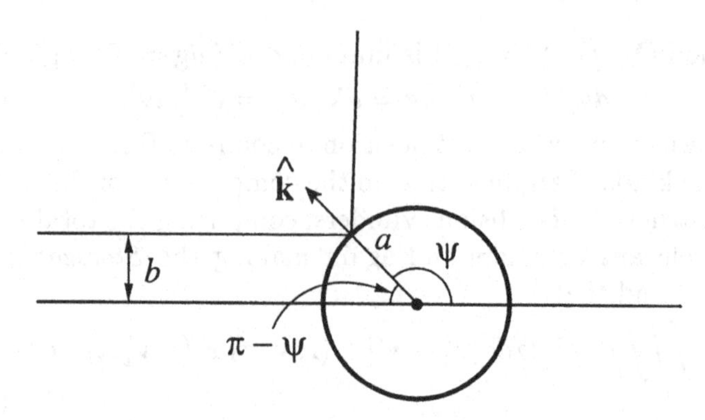

Fig. 2.6 Hard-sphere collision geometry in the plane of the collision. Here, a is the range of the potential, or, equivalently, the diameter of the spheres.

and ψ is the angle between $\mathbf{g}$ and $\hat{\mathbf{k}}$. Then $d\hat{\mathbf{k}} = |\sin\psi\, d\psi\, d\epsilon|$, so that

$$B(\mathbf{g}, \hat{\mathbf{k}}) = |\mathbf{v}_1 - \mathbf{v}| \left|\frac{b}{\sin\psi}\right| \left|\frac{db}{d\psi}\right|, \tag{2.16}$$

with the restriction for purely repulsive potentials that $\mathbf{g} \cdot \hat{\mathbf{k}} < 0$. As can be seen in Fig. 2.4,

$$B(\mathbf{g}', \hat{\mathbf{k}}) = B(\mathbf{g}, -\hat{\mathbf{k}}). \tag{2.17}$$

Let's apply this to the situation where the molecules are hard spheres of *diameter* a. We have $db/d\psi = d(a\sin\psi)/d\psi = a\cos\psi$ (see Fig. 2.6), and $B(\mathbf{g}, \hat{\mathbf{k}}) = ga^2\cos\psi = a^2|(\mathbf{g}\cdot\hat{\mathbf{k}})|$.

The Boltzmann equation for hard spheres is given then as

$$\frac{\partial F}{\partial t} + (\mathbf{v}\cdot\nabla F) = a^2 \int d\mathbf{v}_1 \int_{\mathbf{g}\cdot\hat{\mathbf{k}}<0} d\hat{\mathbf{k}} \left|\mathbf{g}\cdot\hat{\mathbf{k}}\right| \left[F_1' F' - F_1 F\right]. \tag{2.18}$$

This completes the heuristic derivation of the Boltzmann transport equation. Now we turn to Boltzmann's argument that his equation implies the Clausius form of the second law of thermodynamics, namely, that the entropy of an isolated system will increase as the result of any irreversible process taking place in the system. This result is referred to as *Boltzmann's H-theorem*.

2.2 Boltzmann's H-theorem

Boltzmann showed that under very general circumstances, there exists a time-dependent quantity, $H(t)$, that never increases in the course of time. This quantity is given by

$$H(t) \equiv \int d\mathbf{r}_1 \int d\mathbf{v}_1 F(\mathbf{r}_1, \mathbf{v}_1, t) \left[\ln F(\mathbf{r}_1, \mathbf{v}_1, t) - 1\right]. \qquad (2.19)$$

Here, the spatial integral is to be carried out over the entire volume of the vessel containing the gas, and for convenience we have changed the notation slightly. Now we differentiate H with time,

$$\frac{dH(t)}{dt} = \int d\mathbf{r}_1 \int d\mathbf{v}_1 \frac{\partial F_1}{\partial t} \ln F_1, \qquad (2.20)$$

and use the Boltzmann equation to find that

$$\frac{dH}{dt} = \iint d\mathbf{r}_1\, d\mathbf{v}_1 \left[-(\mathbf{v}_1 \cdot \nabla F_1) + J(F_1, F_1) + T_{\mathrm{w}}\right] \ln F_1. \quad (2.21)$$

We are going to carry out some spatial integrations here. We suppose that the distribution function vanishes at the surface of the container and that there is no flow of energy or momentum into or out of the container. (We mention in passing that it is possible to relax this latter condition and thereby obtain a more general form of the second law than we discuss here. This requires a careful analysis of the wall-collision term T_{w}. The interested reader is referred to the article by Dorfman and van Beijeren. Here, we will drop the wall operator since for the purposes of this discussion it merely ensures that the distribution function vanishes at the surface of the container). The first term can be written as

$$-\iint d\mathbf{r}_1\, d\mathbf{v}_1\, \mathbf{v}_1 \cdot \nabla \left[F_1 \left(\ln F_1 - 1\right)\right]. \qquad (2.22)$$

This can be evaluated easily in terms of the distribution function at the walls of the closed container and therefore it is zero. The second term of (2.21) is based on the *Stosszahlansatz*, and is

$$\frac{dH(t)}{dt} = \iiiint d\mathbf{r}_1\, d\mathbf{v}_1\, d\mathbf{v}_2\, d\hat{\mathbf{k}}\; B(\mathbf{g}, \hat{\mathbf{k}})\, \Psi(\mathbf{v}_1)(F_1' F_2' - F_1 F_2), \qquad (2.23)$$

with $\Psi(\mathbf{v}_1) = \ln F_1$. The integrand may be symmetrized in $\mathbf{v}_1$ and $\mathbf{v}_2$ to give

$$\begin{aligned}\frac{dH(t)}{dt} &= \frac{1}{2} \iiiint d\mathbf{r}_1\, d\mathbf{v}_1\, d\mathbf{v}_2\, d\hat{\mathbf{k}}\; B(\mathbf{g}, \hat{\mathbf{k}}) \times \\ &\qquad \left[\Psi(\mathbf{v}_1) + \Psi(\mathbf{v}_2)\right] \left(F_1' F_2' - F_1 F_2\right).\end{aligned}$$

For each collision there is an inverse one, so we can also express the time derivative of the H-function in terms of the inverse collisions as

$$\begin{aligned}\frac{dH(t)}{dt} &= +\frac{1}{2}\iiiint d\mathbf{r}_1\, d\mathbf{v}_1'\, d\mathbf{v}_2'\, d\hat{\mathbf{k}}\; B(\mathbf{g}', -\hat{\mathbf{k}}) \times \\ &\qquad [\Psi(\mathbf{v}_1') + \Psi(\mathbf{v}_2')]\,(F_1F_2 - F_1'F_2') \\ &= -\frac{1}{2}\iiiint d\mathbf{r}_1\, d\mathbf{v}_1\, d\mathbf{v}_2\, d\hat{\mathbf{k}}\; B(\mathbf{g}, \hat{\mathbf{k}}) \times \\ &\qquad [\Psi(\mathbf{v}_1') + \Psi(\mathbf{v}_2')]\,(F_1'F_2' - F_1F_2).\end{aligned}$$

We obtain the H-theorem by adding these expressions and dividing by 2,

$$\begin{aligned}\frac{dH(t)}{dt} &= \frac{1}{4}\iiiint d\mathbf{r}_1\, d\mathbf{v}_1\, d\mathbf{v}_2\, d\hat{\mathbf{k}}\; B(\mathbf{g}, \hat{\mathbf{k}}) \times \\ &\qquad [\Psi_1 + \Psi_2 - \Psi_1' - \Psi_2']\,(F_1'F_2' - F_1F_2).\end{aligned}$$

Now, using $\Psi(\mathbf{v}_1) = \ln F_1$, we obtain

$$\begin{aligned}\frac{dH}{dt} &= \frac{1}{4}\iiiint d\mathbf{r}_1\, d\mathbf{v}_1\, d\mathbf{v}_2\, d\hat{\mathbf{k}}\; B(\mathbf{g}, \hat{\mathbf{k}}) \times \\ &\qquad \ln\left(\frac{F_1F_2}{F_1'F_2'}\right)(F_1'F_2' - F_1F_2).\end{aligned}$$

If $F_1F_2 \neq F_1'F_2'$, the integrand is negative;

$F_1F_2 < F_1'F_2'$: the second factor is positive, the first is negative;
$F_1F_2 > F_1'F_2'$: the second is negative, and the first is positive.

Both cases give a decreasing $H(t)$. That is,

$$\frac{dH(t)}{dt} \leq 0. \tag{2.24}$$

The integral is zero only if for all $\mathbf{v}_1$ and $\mathbf{v}_2$

$$F_1F_2 = F_1'F_2'. \tag{2.25}$$

This is Boltzmann's H-theorem.

We now show that when H is constant in time, the gas is in equilibrium. The existence of an equilibrium state requires the rates of the restituting and direct collisions to be equal; that is, that there is a detailed balance of gain and loss processes taking place in the gas.

Taking the natural logarithm of (2.25), we see that $\ln F_1 + \ln F_2$ has to be conserved for an equilibrium solution of the Boltzmann equation. Therefore, $\ln F_1$ can generally be expressed as a linear combination with constant coefficients of the $(d + 2)$ quantities conserved by binary collisions, i.e, (i) the number of particles, (ii) the d components of the linear momentum, where d is the number of dimensions, and (iii) the kinetic energy: $\ln F_1 = A + \mathbf{B} \cdot \mathbf{v}_1 + Cv_1^2$. (Adding an angular momentum term to $\ln(F_1F_2)$ is not independent of conservation of momentum, because the positions of the particles are the same.) The particles are assumed to have no internal degrees of freedom.

Then $F_1 \propto \exp(\mathbf{B}\cdot\mathbf{v}_1 + C\mathbf{v}_1^2) = A\exp[-\frac{1}{2}\beta m(\mathbf{v}_1 - \mathbf{u})^2]$. When H has reached its minimum value this is the well known Maxwell–Boltzmann distribution for a gas in thermal equilibrium with a uniform motion $\mathbf{u}$. So, argues Boltzmann, solutions of his equation for an isolated system approach an equilibrium state, just as real gases seem to do.

Up to a negative factor ($-k_B$, in fact), differences in H are the same as differences in the thermodynamic entropy between initial and final equilibrium states. Boltzmann thought that his H-theorem gave a foundation of the second law of thermodynamics. The increase in entropy is a result of the collision integral, whose derivation was based on the *Stosszahlansatz.*

This result raises a large number of questions, particularly the central one: Can a gas that is described exactly by the reversible laws of mechanics be characterized by a quantity that always decreases? Perhaps a *nonmechanical* assumption was introduced here. If so, this would suggest, although not imply, that Boltzmann's equation might not be a useful description of nature. In fact, though, this equation is so useful and accurate a predictor of the properties of dilute gases, that it is now often used as a test of intermolecular potential models.

Our next step is to see if we can understand the role of the *Stosszahlansatz* in Boltzmann's derivation and to see if we can understand on an intuitive level how to reconcile irreversible behavior with mechanical reversibility of the underlying dynamics. Also we need to understand, if possible, why the Boltzmann equation works so well. To develop answers to these questions, we first turn to a simple model introduced by Mark Kac which sheds some light on them.

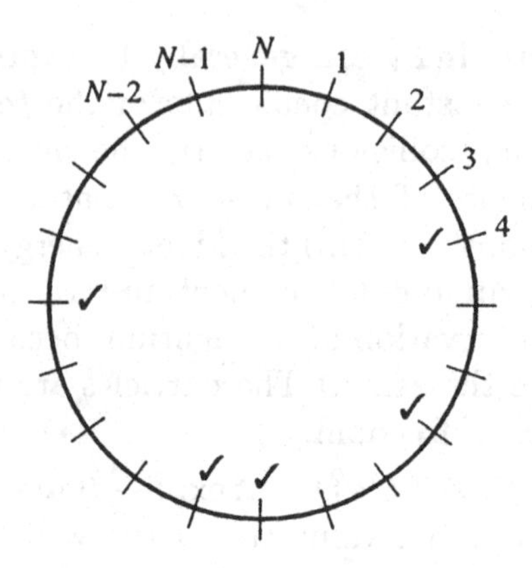

Fig. 2.7. The Kac ring model.

2.3 Kac's ring model

The Kac ring model doesn't look much like the system of particles colliding with each other considered by Boltzmann. However, like many clever models developed to help understand complex phenomena, this model is of great didactic value, and clarifies many issues. It does have a kind of 'collision' and the *Stosszahlansatz* can be studied in detail from a more microscopic point of view. We proceed as follows.

N lattice points are placed uniformly around a circle so that it is subdivided into N segments. Each segment contains one ball that can either be black or white. The motion is discretized into a sequence of time steps. At each time step all of the balls shift one place clockwise (see Fig. 2.7). On *some* of the lattice sites there is a fixed marker whose job is to change the color of every ball that passes it. We wish to obtain an equation for the number of balls of each color after every time step.

Define the following quantities:

$B(t)$ $\equiv$ the number of black balls at time t.
$W(t)$ $\equiv$ the number of white balls at time t.
$b(t)$ $\equiv$ the number of black balls in front of a marker at time t.
$w(t)$ $\equiv$ the number of white balls in front of a marker at time t.

It then follows that

$$B(t+1) \quad = \quad B(t) + w(t) - b(t),$$

$$W(t+1) = W(t) + b(t) - w(t). \tag{2.26}$$

The difference between the number of black and white balls, $\Delta(t)$, satisfies the equation

$$\begin{aligned} \Delta(t+1) &= B(t+1) - W(t+1) \\ &= \Delta(t) - 2\left[b(t) - w(t)\right]. \end{aligned} \tag{2.27}$$

This equation is exact, so far. However, there is little we can do with it. In order to calculate $\Delta(t)$, we can make an assumption similar to the *Stosszahlansatz.* We suppose that the fraction of white or black balls that change color at a given time step is equal to the probability μ that a lattice site has a marker on it, where μ is equal to the number of markers divided by the number of sites, N. That is,

$$\frac{b(t)}{B(t)} = \frac{w(t)}{W(t)} = \mu. \tag{2.28}$$

Now we can solve (2.27), because it becomes

$$\Delta(t+1) = \Delta(t) - 2\mu\left(B(t) - W(t)\right) \tag{2.29}$$

$$= (1-2\mu)\Delta(t). \tag{2.30}$$

This can be iterated for the first t time steps, and we obtain

$$\Delta(t) = \Delta(0)\,(1-2\mu)^t. \tag{2.31}$$

Since $\mu \leq 1$, (2.31) predicts a strictly monotonic decay for $|\Delta\,(t)| \to 0$ in the limit that $t \to \infty$. This solution cannot be a very good description of the system for *all* times, because it is not consistent with many things that we know must be true:

1. Equation (2.31) is not-time reversible while the elementary dynamics of this model is reversible. A change of sign of t simply changes a monotonically decreasing function into a monotonically increasing one. By looking at some simple examples, one can easily convince oneself that this equation cannot always be correct.
2. There is always a number of steps, $t = 2N$, when the initial state of the system must be recovered. This is clear because after this number of time steps, each ball will have passed each marker twice, so each ball must be at its initial location with its initial color.† This feature is not present in (2.31).

† This is a trivial example of the Poincaré recurrence theorem, to be discussed in Chapter 3.

To resolve the contradiction between the reversible microscopic dynamics and the irreversible (2.31), we need to follow the suggestion of Boltzmann and adopt the point of view that (2.31) refers to the *most probable behavior of a member of an ensemble of systems* rather than to the *exact* behavior of any member of the ensemble. One of the virtues of the Kac ring model is that we can easily see how this suggestion can be implemented. We can also see how both Loschmidt's and Zermelo's paradoxes are resolved in this model.

In this case, the ensemble is a large collection of Kac's rings that initially have the same distribution of black and white balls, but with different distributions of markers. For each site on the ring labelled by i independently, a fraction μ of the ith site of all members of the ensemble has a marker. The following variables are used to describe the 'microscopic' state of a ring. For each site i, we define

$$\begin{aligned}\eta_i(t) &= +1 \text{ if there is a black ball in front of site } i \text{ at time } t,\\ \eta_i(t) &= -1 \text{ if there is a white ball in front of site } i \text{ at time } t,\\ \epsilon_i &= +1 \text{ if there is no marker at } i,\\ \epsilon_i &= -1 \text{ if there is a marker at } i.\end{aligned}$$

Then, $\Delta(t) = \sum_i \eta_i(t)$ and $\eta_{i+1}(t+1) = \epsilon_i \eta_i(t)$. Using this relation $t+1$ times to express the left-hand side in terms of the initial state of the system, we obtain

$$\eta_{i+1}(t+1) = \epsilon_i \epsilon_{i-1} \dots \epsilon_{i-t} \eta_{i-t}(0). \tag{2.32}$$

Therefore,

$$\Delta(t+1) = \sum_i \epsilon_i \epsilon_{i-1} \dots \epsilon_{i-t} \eta_{i-t}(0). \tag{2.33}$$

This is consistent with the recurrence time, since after $2N$ steps each ϵ appears twice and $\Delta(2N) = \Delta(0)$.

The ensemble average is given by

$$\langle \Delta(t) \rangle_{\text{ens}} = \sum_i \langle \epsilon_{i-1} \epsilon_{i-2} \dots \epsilon_{i-t} \rangle \eta_{i-t}(0), \tag{2.34}$$

where the average on the right-hand side denotes the average of t successive ϵs. The indices can also be taken from 1 to t, since the distribution of markers is uniform over a ring. This means that this average can be taken outside the summation, so that

$$\langle \Delta(t) \rangle = \langle \epsilon_1 \dots \epsilon_t \rangle \, \Delta(0). \tag{2.35}$$

Two cases need to be distinguished: the number of time steps is less than N or it is larger than N. In the first case we don't need to worry about the periodicity of the system. The probability that a ball passes j markers in t steps is $\mu^j(1-\mu)^{t-j}$ multiplied by the number of possible ways to distribute them over t places, which is $t!/j!(t-j)!$. In the case of j markers, $\epsilon_1 \dots \epsilon_t = (-1)^j$. Therefore for $0 < t < N$, we have

$$\langle \epsilon_1 \dots \epsilon_t \rangle = \sum_j^t \frac{t!}{j!(t-j)!}(-1)^j \mu^j (1-\mu)^{t-j} = (1-2\mu)^t. \quad (2.36)$$

This leads to a result identical to (2.31), on the basis of the *Stosszahlansatz*. In particular,

$$\begin{aligned} \langle \Delta(t) \rangle &= \langle \epsilon_1 \dots \epsilon_t \rangle \, \Delta(0) \\ &= (1-2\mu)^t \Delta(0) \quad \text{for } 0 \leq t \leq N. \end{aligned} \quad (2.37)$$

The calculation of the ensemble average given above can be understood by the following argument: we assign a random number between zero and 1 to each bond connecting nearest neighbors on the ring. Then if μ is less than the random number, a marker is placed on the bond, and otherwise, no marker is placed. These random numbers are fixed for all times, so that a given configuration of markers remains fixed on a ring.

In view of this picture of the ensemble, we can now turn to the case where $2N \geq t > N$. Then $t = N + s$, and we express $\Delta(t) = \langle \epsilon_1 \dots \epsilon_{N+s} \rangle$ as

$$\begin{aligned} \langle \epsilon_1 \dots \epsilon_{N+s} \rangle &= \langle \epsilon_1 \epsilon_2 \dots \epsilon_N \epsilon_1 \epsilon_2 \dots \epsilon_{N+s} \rangle = \langle \epsilon_{N+s+1} \dots \epsilon_{2N} \rangle \\ &= \langle \epsilon_{s+1} \dots \epsilon_N \rangle = \langle \epsilon_1 \dots \epsilon_{N-s} \rangle = \langle \epsilon_1 \dots \epsilon_{2N-t} \rangle , \end{aligned}$$

because the markers are distributed uniformly. The expression for Δ is the same as that for $t < N$, but with t replaced by $2N - t$,

$$\langle \Delta(t) \rangle_{\text{ens}} = (1-2\mu)^{2N-t} \Delta(0) \quad \text{for } N < t \leq 2N. \quad (2.38)$$

For $t \to 2N$ the average of this ensemble attains its initial value again.

We see that the Boltzmann equation result, (2.31), is identical with (2.37). Starting then from some initial value $\Delta(0)$ and considering times $t \ll N$, we recover the Boltzmann equation from an ensemble average, after treating the dynamics correctly. The third issue that needs investigating is whether this result is typical, in the sense that almost all members of the ensemble will exhibit the same behavior, at least to within some small experimental accuracy. If N is sufficiently large, we might expect to be able to prove this, since the markers will be distributed more or less randomly in any particular member of the ensemble. That is, the probability of finding large clusters of markers or large regions free of markers, which would lead to non-Boltzmann-like behavior will become exponentially small as N grows large. An analysis of the fluctuations about Boltzmann-like behavior can be found in the book by Kac, listed below, and it verifies this intuitive consideration. It is worth mentioning that there are situations where improbable clusters do have physical effects. These are often connected with the phenomenon called *Lifshitz tails*, and an example is discussed in Sec. 16.3.

Let us turn to the resolution of Zermelo's time-reversal paradox for the moment. Suppose we let all of the systems run for a time t, with $0 \ll t \ll N$, and then time-reverse them. We would see that the Boltzmann equation would give totally wrong results, since the actual evolution would be toward the common initial value $\Delta(0)$. However, the state of each of these systems is highly correlated at time t, in the sense that each of the moving particles have already encountered whatever scatterers are located in the first t sites in front of them. In the backward motion, they will simply retrace their steps and arrive at the earlier initial state. The Boltzmann equation describes the evolution of an initial uncorrelated state, which is statistically likely in our ensemble, while the time-reversed motion describes the evolution from a correlated state, which is statistically a very unlikely initial state in our ensemble, since it would require the colors of the balls and the locations of the markers to match up in such a way that a definite state is produced after t steps. This is exactly what happens when $N < t < 2N$, for at $t = N$, each particle has encountered each scatterer once. This is then a highly correlated state, which leads to 'anti-Boltzmann' behavior over the next N time steps.†

† There are some nice computer simulations of a gas where initially the gas is confined to one half of a vessel, and the gas molecules can reach the other half of the vessel by going through a small hole in a partition separating the two halves. The simulation allows the motion to proceed until the two halves of the container are about equally filled. Then the motion is reversed, and one can watch the reversed motion back to the initial state. However, if, just before the motion is reversed one perturbs the state a little bit, by

Now we can see how the program for statistical mechanics outlined in the previous chapter is carried out in the simple case of the Kac ring model. Note that for this model, as the number of sites gets larger, the time interval over which the Boltzmann equation is valid grows larger, too. However, as the time approaches the recurrence time, the Boltzmann equation is not valid, even for the ensemble average, and it must be replaced by an 'anti-Boltzmann' equation. A correct microscopic treatment of this model shows both the Boltzmann-like as well as the anti-Boltzmann-like behavior. We also note that the choice of the ensemble, with the random distribution of markers, is an essential feature of the model and this assists, so to speak, the Boltzmann-like evolution of the quantity $\langle\Delta\rangle$. When we consider more realistic dynamical systems in later chapters, we will look for signs that the microscopic dynamics itself, while mechanically reversible, nevertheless provides a stochastic-like element that is responsible for the approach to equilibrium.

2.4 Tagged particle diffusion

It is one thing to derive the Boltzmann equation and to discuss the fundamental assumptions in the derivation. It is quite another thing to solve this equation. This is an entire subject in its own right. Here, we illustrate the standard Chapman–Enskog method of solution for a simple nonequilibrium process in a fluid, namely, tagged particle diffusion. We imagine that a particle is immersed in a dilute gas that is in equilibrium. The immersed particle is mechanically identical to all of the other particles in the fluid, but it is tagged in some way that does not disturb its mechanical properties. On a computer this is easily accomplished, but in a real system one would have to use small isotopic differences as a tag on the particle. Our object is to determine the distribution function for the tagged particle. To simplify matters still further, we will consider a hypothetical two-dimensional system. This only

adding a few extra particles, say, then the initial is not recovered, and the density in the second half never reaches zero in the time-reversed motion. This is a clear example of the fact that a time-reversed state is unstable and a small perturbation will not behave in the same way. I thank Prof. M. Droz, University of Geneva, for this example.

avoids some minor technicalities with three-dimensional vectors, which are not interesting here.

We suppose that the tagged particle collides with gas particles whose distribution function is an equilibrium Maxwell–Boltzmann function $F_{\mathrm{b}}(\mathbf{v}) = n\phi_0(\mathbf{v})$, where the subscript b denotes the untagged particles which provide a 'bath' for the tagged particle,

$$\phi_0(\mathbf{v}) = \left(\frac{\beta m}{2\pi}\right)^{d/2} e^{-\frac{1}{2}\beta m v^2}, \tag{2.39}$$

n is the number density of the bath-particles and d is the number of space dimensions. The Boltzmann equation for the tagged particle is

$$\frac{\partial F_{\mathrm{T}}(\mathbf{r},\mathbf{v},t)}{\partial t} + \mathbf{v}\cdot\nabla F_{\mathrm{T}}(\mathbf{r},\mathbf{v},t) = \int d\mathbf{v}_1 \int d\hat{\mathbf{k}} B(\mathbf{g},\hat{\mathbf{k}})\left[F'_{\mathrm{b}}F'_{\mathrm{T}} - F_{\mathrm{b}}F_{\mathrm{T}}\right]. \tag{2.40}$$

The probability of finding the tagged particle at point $\mathbf{r}$ at time t regardless of its velocity is given by $P(\mathbf{r},t)$, where

$$P(\mathbf{r},t) = \int d\mathbf{v}\, F_{\mathrm{T}}(\mathbf{r},\mathbf{v},t). \tag{2.41}$$

We expect this probability to satisfy a diffusion equation of the form

$$\frac{\partial P(\mathbf{r},t)}{\partial t} = D\nabla^2 P(\mathbf{r},t), \tag{2.42}$$

where D, the diffusion coefficient, is now to be determined.

We begin our derivation of this diffusion equation, and the determination of the diffusion coefficient, by integrating the Boltzmann equation, (2.40), over the velocity of the tagged particle, to obtain

$$\frac{\partial P(\mathbf{r},t)}{\partial t} + \int d\mathbf{v}\,[\mathbf{v}\cdot\nabla F_{\mathrm{T}}(\mathbf{r},\mathbf{v},t)] = \int d\mathbf{v}\int d\mathbf{v}_1 \int d\hat{\mathbf{k}}\, B(\mathbf{g},\hat{\mathbf{k}})[F'_{\mathrm{T}}F'_{\mathrm{b}} - F_{\mathrm{T}}F_{\mathrm{b}}]. \tag{2.43}$$

The right-hand side of this equation is zero because the primed and unprimed variables can be interchanged as in the derivation of the H-theorem, since $d\mathbf{v}_1 d\mathbf{v} = d\mathbf{v}'_1 d\mathbf{v}'$, and $B(\mathbf{g},\hat{\mathbf{k}}) = B(\mathbf{g}',-\hat{\mathbf{k}})$. This yields a conservation equation for the probability density for the tagged particle

$$\frac{\partial P(\mathbf{r},t)}{\partial t} + \nabla\cdot\mathbf{J}_{\mathrm{T}} = 0, \tag{2.44}$$

where the probability current for the tagged particle, $\mathbf{J}_{\mathrm{T}}$, is

$$\mathbf{J}_{\mathrm{T}}(\mathbf{r},t)=\int d\mathbf{v}\,\mathbf{v}\,F_{\mathrm{T}}(\mathbf{r},\mathbf{v},t). \tag{2.45}$$

The essence of the Chapman–Enskog method for the solution of (2.40) is to note that if F_{T} were to have the form of a Maxwell–Boltzmann velocity distribution, then the collision integral on the right-hand side of (2.41) would vanish, and F_{T} would only change in time due to the free streaming of the tagged particle in the gas. This suggests the following picture for the change in F_{T}. The tagged particle is injected in the gas at some initial time. There is a 'kinetic' stage where the distribution function for the tagged particle changes rapidly in time due to collisions. The kinetic stage lasts for a few mean free times or so. After this period, velocity distribution for the tagged particle is almost, but not quite, the same as that for other particles in the gas, namely a Maxwell–Boltzmann distribution. From then on, the spatial probability distribution for the tagged particle slowly becomes uniform throughout the container, as the tagged particle diffuses. This last stage is a 'hydrodynamic' stage, largely governed by diffusion of the particle in the gas.

This picture can be expressed mathematically by supposing that, in the hydrodynamic stage, F_{T} may be expanded about its 'local' equilibrium form in a power series in the parameter $\mu=l/L$, where l is the mean free path length, and L is the macroscopic distance over which the probability density varies. In the limit $L\to\infty$, the system is homogeneous, $\mu=0$, and the system is in equilibrium. Then

$$F_{\mathrm{T}}(\mathbf{r},\mathbf{v},t)=P(\mathbf{r},t)\phi_0(\mathbf{v})\left[1+\mu\Phi_1(\mathbf{r},\mathbf{v},t)+\mu^2\Phi_2(\mathbf{r},\mathbf{v},t)+\cdots\right], \tag{2.46}$$

where the functions $\Phi_1,\Phi_2,\ldots$ are to be determined. It is assumed that the $\mu^j\Phi_j$ are proportional to the jth gradient of $P(\mathbf{r},t)$ so that (2.46) may be considered to be a systematic expansion of F_{T} in powers of the gradients. There is a consistency condition that the Φ_i must satisfy, coming from (2.41) and (2.46), namely that

$$\int d\mathbf{v}\phi_0(\mathbf{v})\Phi_i(\mathbf{r},\mathbf{v},t)=0. \tag{2.47}$$

When this expansion is inserted in the conservation law, (2.44), one obtains

$$\frac{\partial P(\mathbf{r},t)}{\partial t} = -\mu\nabla\cdot\int d\mathbf{v}\,\mathbf{v}P(\mathbf{r},t)\phi_0(\mathbf{v})\left[1+\mu\Phi_1+\mu^2\Phi_2+\cdots\right]. \tag{2.48}$$

We have inserted a factor of μ in front of the differential operator to keep track of the fact that differentiation produces a factor proportional to the gradient of P.† The first term on the right-hand side is odd in $\mathbf{v}$, so it is zero. The other terms are at least of second order in the gradients. So the expansion is consistent with the conservation law if $\partial P/\partial t$ and $\nabla^2 P$ are of equal order, that is, they both are formally of order μ^2.

To first order, the tagged particle probability current,

$$\mathbf{J}_\mathrm{T} = \mu\int d\mathbf{v}\,\mathbf{v}P(\mathbf{r},t)\phi_0(\mathbf{v})\Phi_1(\mathbf{r},\mathbf{v},t)+\cdots, \tag{2.49}$$

is proportional to ∇P because that is the only available vector in the integrand. This proportionality constant, D, is the diffusion coefficient that appears in the diffusion equation

$$\frac{\partial P}{\partial t} = \nabla\cdot D\nabla P(\mathbf{r},t).$$

The expansion of the distribution function given by (2.46) is now inserted in the Boltzmann equation:

$$\begin{aligned}\left(\mu^2\frac{\partial}{\partial t}+\mu\mathbf{v}\cdot\nabla\right)&[P(\mathbf{r},t)\phi_0(\mathbf{v})(1+\mu\Phi_1+\cdots)]\\ = n\mu\int d\mathbf{v}_1\int_{\mathbf{g}\cdot\hat{\mathbf{k}}<0} & d\hat{\mathbf{k}}B(\mathbf{g},\hat{\mathbf{k}})P(\mathbf{r},t)\phi_0(\mathbf{v})\phi_0(\mathbf{v}_1)\times\\ &[\Phi_1(\mathbf{v}')-\Phi_1(\mathbf{v})]+\cdots.\end{aligned} \tag{2.50}$$

Here, we have included powers of μ to indicate explicitly that the time derivative of P is of order μ^2, and that its spatial derivative is of order μ. Collecting all terms of first order in μ, we obtain

$$\begin{aligned}&\phi_0(\mathbf{v})\mathbf{v}\cdot\nabla P(\mathbf{r},t)\\ &\quad = nP(\mathbf{r},t)\int d\mathbf{v}_1\int_{\mathbf{g}\cdot\hat{\mathbf{k}}<0} d\hat{\mathbf{k}}\,B(\mathbf{g},\hat{\mathbf{k}})\phi_0(\mathbf{v}_1)\phi_0(\mathbf{v})\times\\ &\qquad\qquad [\Phi_1(\mathbf{v}')-\Phi_1(\mathbf{v})].\end{aligned} \tag{2.51}$$

This is a linear, inhomogeneous equation in Φ_1 of the general form of a linear operator $\mathcal{L}$ acting on the function Φ_1, $\mathcal{L}\Phi_1 = \alpha(v)$, with

† This is just a *trick* to enable us to keep track of the various orders in l/L. We can set $\mu = 1$ at the end.

an inhomogeneous term. A solution would take the form

$$\Phi_1 = \mathcal{L}^{-1}\alpha(v) + \Phi_1^0,$$

where Φ_1^0 is a general solution of the homogeneous equation, $\mathcal{L}\Phi_1^0 = 0$, provided the inhomogeneous term, $\alpha(v)$, is orthogonal to Φ_1^0. It is clear from (2.51) that Φ_1^0 must be proportional to $\phi_0(\mathbf{v})$ and that the inhomogeneous term, the left-hand side of (2.51), is in fact orthogonal to it. Moreover, using (2.47), we can see that Φ_1^0 must be set equal to zero.

We expand $\Phi_1(\mathbf{v})$ in a complete set of orthogonal polynomials, the Sonine polynomials. These polynomials are often used to construct solutions to the Boltzmann equation. They are exact eigenfunctions of the linearized Boltzmann collision operator for Maxwell molecules which interact with repulsive r^{-4} potentials, and they lead to rapidly converging solutions of the linearized Boltzmann equation for other molecules that interact with short-range repulsive potentials. The use of the Sonine polynomials is discussed in great detail by Chapman and Cowling in their classic book on kinetic theory, listed in the references below.

For the purpose of illustration, we will keep only the first term in an approximate solution of (2.51) in terms of Sonine polynomials. That is, we set $\Phi_1(\mathbf{v}) = \mathbf{a}\cdot\mathbf{v}$, where the vector $\mathbf{a}$ is taken to be a function of position but not of velocity. We must now determine $\mathbf{a}$. If we insert this approximate solution in (2.51), multiply by $\mathbf{v}$ and then integrate over $\mathbf{v}$, we find

$$\begin{aligned}\int d\mathbf{v}\phi_0(\mathbf{v})\mathbf{v}(\mathbf{v}\cdot\nabla P) &= \\ nP\int d\mathbf{v}_1\int d\mathbf{v}\int d\hat{\mathbf{k}}\; B(\mathbf{g},\hat{\mathbf{k}})\;\phi_0(\mathbf{v})\;\phi_0(\mathbf{v}_1)\,\times& \\ \mathbf{v}(\mathbf{a}\cdot\mathbf{v}' - \mathbf{a}\cdot\mathbf{v}).& \qquad (2.52)\end{aligned}$$

By considering only the x-component, we find that the left-hand side of (2.51) becomes

$$\text{LHS}=\int d\mathbf{v}\phi_0(\mathbf{v})v_x(\mathbf{v}\cdot\nabla P) = \frac{\partial P}{\partial x}\int d\mathbf{v}\phi_0(\mathbf{v})v_x^2 = \frac{1}{\beta m}\frac{\partial P}{\partial x},$$

because the integrand is odd in v_y.

The particles are assumed to be hard two-dimensional discs with diameter a. Then $B(\mathbf{g},\hat{\mathbf{k}}) = a|\mathbf{g}\cdot\hat{\mathbf{k}}|$ (see Fig. 2.6), so the

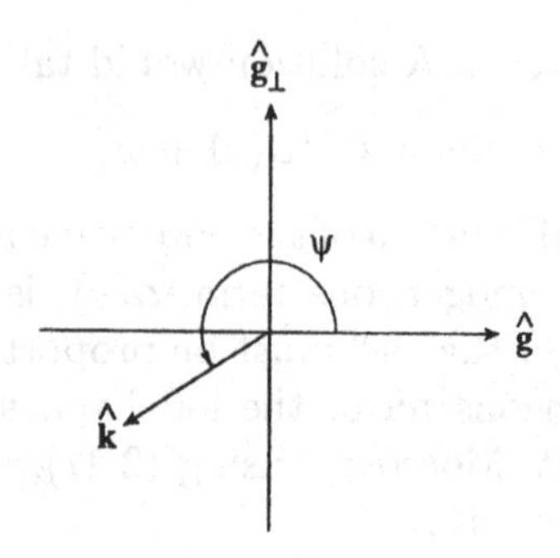

Fig. 2.8. The geometry for the $\hat{\mathbf{k}}$-integration.

right-hand side of the x-component of (2.51) becomes

$$\text{RHS} = naP\int d\mathbf{v}\int d\mathbf{v}_1\int_{\mathbf{g}\cdot\hat{\mathbf{k}}<0} d\psi|\mathbf{g}||\cos(\psi)|\phi_0(\mathbf{v})\phi_0(\mathbf{v}_1)\times v_x(\mathbf{a}\cdot\hat{\mathbf{k}})(\mathbf{g}\cdot\hat{\mathbf{k}}). \tag{2.53}$$

Here, $\mathbf{v}'-\mathbf{v}=(\mathbf{g}\cdot\hat{\mathbf{k}})\hat{\mathbf{k}}$ is used.

We use a simple decomposition of the vector $\hat{\mathbf{k}}$ in directions parallel and perpendicular to $\mathbf{g}$. That is,

$$\hat{\mathbf{k}} = \left[\left(\hat{\mathbf{g}}\cdot\hat{\mathbf{k}}\right)\hat{\mathbf{g}} + \left(\hat{\mathbf{g}}_\perp\cdot\hat{\mathbf{k}}\right)\hat{\mathbf{g}}_\perp\right], \tag{2.54}$$

where the unit vectors $\hat{\mathbf{g}}$, $\hat{\mathbf{g}}_\perp$ are parallel or perpendicular, respectively, to $\mathbf{g}$. Then the integral over ψ (see Fig. 2.8) becomes

$$\begin{aligned}
&\int_{\frac{1}{2}\pi}^{\frac{3}{2}\pi} d\psi|\cos(\psi)|\left[\left(\hat{\mathbf{g}}\cdot\hat{\mathbf{k}}\right)(\mathbf{a}\cdot\hat{\mathbf{g}}) + \left(\hat{\mathbf{g}}_\perp\cdot\hat{\mathbf{k}}\right)(\mathbf{a}\cdot\hat{\mathbf{g}}_\perp)\right]\left(\hat{\mathbf{g}}\cdot\hat{\mathbf{k}}\right)\\
&= \int_{\frac{1}{2}\pi}^{\frac{3}{2}\pi} d\psi|\cos(\psi)|\left[(\mathbf{a}\cdot\hat{\mathbf{g}})\cos(\psi) + (\mathbf{a}\cdot\hat{\mathbf{g}}_\perp)\sin(\psi)\right]\cos(\psi)\\
&= \frac{4}{3}(\mathbf{a}\cdot\hat{\mathbf{g}}).
\end{aligned} \tag{2.55}$$

Now we have

$$\text{RHS} = naP\frac{4}{3}\int d\mathbf{v}\int d\mathbf{v}_1|\mathbf{g}|v_x(\mathbf{a}\cdot\mathbf{g})\phi_0(\mathbf{v})\phi_0(\mathbf{v}_1).$$

To do the integral over $\mathbf{v}$ and $\mathbf{v}_1$, we convert to center of mass and relative velocities, $\mathbf{V}=\frac{1}{2}(\mathbf{v}+\mathbf{v}_1)$ and $\mathbf{g}=\mathbf{v}_1-\mathbf{v}$. Using $d\mathbf{v}_1 d\mathbf{v} = d\mathbf{V}d\mathbf{g}$ and $v_1^2+v_2^2 = 2V^2+\frac{1}{2}g^2$, we find that the integral,

RHS, becomes

$$\begin{aligned}\text{RHS} &= \frac{4}{3}naP\left(\frac{\beta m}{2\pi}\right)^2 \int d\mathbf{V} \int d\mathbf{g}\, e^{-\frac{1}{2}\beta m(2V^2+\frac{1}{2}g^2)} \times \\ &\qquad |\mathbf{g}|(\mathbf{a}\cdot\mathbf{g})\left[V_x - \frac{1}{2}g_x\right].\end{aligned}$$

The first term is odd in V_x. In the inner product, the terms that contain g_y terms are also odd, so that

$$\begin{aligned}\text{RHS} &= -\frac{2}{6}naP\frac{\beta m}{2\pi}\int d\mathbf{g}\, e^{-\frac{1}{4}\beta m g^2}|\mathbf{g}|a_x g_x^2 \\ &= -\frac{a_x naP\beta m}{6\pi}\int_0^\infty g dg d\phi\, e^{-\frac{1}{4}\beta m g^2} g^3\cos^2(\phi) \\ &= -\frac{a_x naP\beta m}{12\pi}\frac{2\pi}{2}\frac{\partial^2}{\partial\beta^2}\frac{16}{m^2}\left(\frac{4\pi}{\beta m}\right)^{\frac{1}{2}} \\ &= -\frac{2naP\beta a_x \pi^{\frac{1}{2}}}{m^{\frac{3}{2}}}\beta^{-\frac{5}{2}}\end{aligned}$$

Setting this equal to the left-hand side of (2.51), we obtain for a_x:

$$a_x = -\frac{1}{2naP}\frac{\partial P}{\partial x}\left(\frac{\beta m}{\pi}\right)^{\frac{1}{2}}. \tag{2.56}$$

Then similar arguments can be made for the other component of $\mathbf{a}$, and we have, to first order in the gradient of P,

$$F_{\mathrm{T}} = P(\mathbf{r},t)\phi_0(\mathbf{v})[1+(\mathbf{a}\cdot\mathbf{v})] = P\phi_0 - \phi_0\frac{1}{2na}\left(\frac{\beta m}{\pi}\right)^{\frac{1}{2}}\left(\mathbf{v}\cdot\frac{\partial P}{\partial \mathbf{r}}\right). \tag{2.57}$$

Substituting this in (2.49), we find that $\mathbf{J}_{\mathrm{T}}$ is

$$\begin{aligned}\mathbf{J}_{\mathrm{T}}(\mathbf{r},t) &= \int d\mathbf{v}\mathbf{v}\phi_0(\mathbf{v})\left[P(\mathbf{r},t) - \frac{1}{2na}\left(\frac{\beta m}{\pi}\right)^{\frac{1}{2}}\left(\mathbf{v}\cdot\frac{\partial P}{\partial \mathbf{r}}\right)\right] \\ &= D\nabla P(\mathbf{r},t),\end{aligned} \tag{2.58}$$

since the first term under the integral is odd in $\mathbf{v}$. The conservation law now is

$$\frac{\partial P(\mathbf{r},t)}{\partial t} = -\nabla\cdot\mathbf{J}_{\mathrm{T}}(\mathbf{r},t) = \nabla\cdot(D\nabla P(\mathbf{r},t)), \tag{2.59}$$

where the diffusion coefficient is given by

$$D = \frac{1}{2na(\pi\beta m)^{\frac{1}{2}}}.$$

This is the usual diffusion equation for an isotropic system,

$$\frac{\partial P(\mathbf{r},t)}{\partial t} = D\nabla^2 P(\mathbf{r},t). \tag{2.60}$$

It is interesting to conclude this section with a brief discussion of some of the properties of the solution of the diffusion equation. We will need these properties later on. We consider a solution of a diffusion equation with the initial condition

$$P(\mathbf{r},t=0) = \delta(\mathbf{r}-\mathbf{r}_0).$$

If we express $P(\mathbf{r},t)$ as a Fourier transform in d-dimensions,

$$P(\mathbf{r},t) = \int \frac{d\mathbf{k}}{(2\pi)^d} e^{i\mathbf{k}\cdot\mathbf{r}} P_k(t), \tag{2.61}$$

and for the δ-function, use

$$P(\mathbf{r},0) = \delta(\mathbf{r}-\mathbf{r}_0) = \int \frac{d\mathbf{k}}{(2\pi)^d} e^{i\mathbf{k}\cdot(\mathbf{r}-\mathbf{r}_0)} = \int \frac{d\mathbf{k}}{(2\pi)^d} P_k(0) e^{i\mathbf{k}\cdot\mathbf{r}},$$

then $P_k(t=0) = e^{-i\mathbf{k}\cdot\mathbf{r}_0}$, and we consider the diffusion equation in an arbitrary number of dimensions, d. The diffusion equation for the Fourier transform is

$$\int \frac{d\mathbf{k}}{(2\pi)^d} \left[\frac{\partial P_k(t)}{\partial t} - Dk^2 P_k(t)\right] e^{i\mathbf{k}\cdot\mathbf{r}} = 0. \tag{2.62}$$

This equation can be solved as

$$P_k(t) = e^{-Dk^2 t} P_k(0) = e^{-Dk^2 t} e^{-i\mathbf{k}\cdot\mathbf{r}_0}. \tag{2.63}$$

Inserting this in (2.61), we obtain

$$P(\mathbf{r},t) = \int \frac{d\mathbf{k}}{(2\pi)^d} e^{-Dk^2 t + i\mathbf{k}\cdot(\mathbf{r}-\mathbf{r}_0)} = \left(\frac{1}{4\pi Dt}\right)^{\frac{1}{2}d} e^{-(\mathbf{r}-\mathbf{r}_0)^2/4Dt}.$$

One of the most important quantities is the mean square displacement of the diffusing particle, which is given as

$$\left\langle (\mathbf{r}-\mathbf{r}_0)^2 \right\rangle_t = \int d\mathbf{r} P(\mathbf{r},t)(\mathbf{r}-\mathbf{r}_0)^2 = 2tDd. \tag{2.64}$$

This linear growth of the *mean square* displacement with time is typical for diffusive motion. It is in contrast to free motion, where $(\mathbf{r}-\mathbf{r}_0)^2 \propto t^2$. Finally, we note the Einstein relation between the mean square displacement and the diffusion coefficient given by

$$D = \frac{\langle (\mathbf{r}-\mathbf{r}_0)^2 \rangle}{2td}. \tag{2.65}$$

This completes our discussion of the Boltzmann equation and various issues related to it. Now we can begin to consider the theory of irreversible processes in fluids from a more fundamental point of view in order to bring out the features of chaotic dynamics which are central to this book. The background just presented is essential for an understanding of the point of our further discussions.

2.5 Further reading

The book of Boltzmann [Bol95] is the inspiration for all later treatments of the subject. Much of the material of this chapter is taken from the review article by Dorfman and van Beijeren [DvB77]. The classic text on the kinetic theory of dilute gases is by Chapman and Cowling [CC70], and more recent texts are by Resibois and de Leener [RdL77], by McLennan [McL89], and by Ferziger and Kaper [FK72], among others. The Kac ring model is described in the article by Kac in the collection edited by Wax [Wax54], in Kac's book [Kac76], and also in the books of Wannier [Wan87] and of Schulman [Sch97]. A good description of the method for setting up the ensemble used in the averaging can be found in the papers of Wilkenson *et al.* and of Bagnoli *et al.*[WW83, BPR97].† Some of the interesting issues related to time-reversibility that we have discussed in this chapter are studied by means of molecular dynamics in a recent paper by Levesque and Verlet [LV93]. Very nice discussions of the Boltzmann equation and related topics are given in the texts by Uhlenbeck and Ford [UF63], and by Kac [Kac76]. Of course, the Encyclopedia article of the Ehrenfests [PE59] should be consulted for a very careful discussion of the *Stosszahlansatz* and related issues. Don't forget to read the footnotes!

2.6 Exercises

1. **The Ehrenfest wind-tree model.** A collection of fixed scatterers ('trees') are placed on a plane at random. The trees are oriented squares with diagonals along the x- and y-directions (see

† I thank Prof. R. Rechtman for helpful discussions, and for these references.

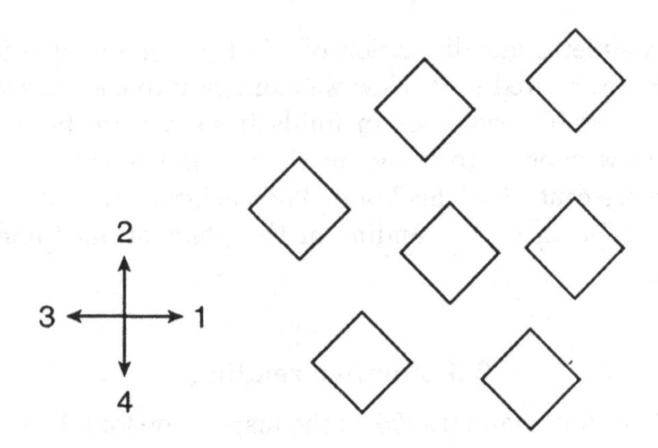

Fig. 2.9. The Ehrenfest wind-tree model.

Fig. 2.9). The number of trees per unit volume is n, each side is of length a, and $na^2 \ll 1$. There are moving particles ('wind') that do not interact with each other, but do collide with the trees. The wind particles can move in four directions, labeled 1, 2, 3, 4. Let $F_i(\mathbf{r}, t) =$ the number of wind particles at $\mathbf{r}$ moving in direction i at time t.

(a) Derive an equation for $F_i(\mathbf{r}, t)$.
(b) Is there an H-theorem? (Suppose the system is spatially homogeneous, $F_i(\mathbf{r}, t) = F_i(t)$, independent of $\mathbf{r}$.)
(c) Find a solution $\{F_i(t)\}$ in terms of $\{F_i(0)\}$. What happens if $t \to \infty$? (You will need to diagonalize a 4×4 matrix).

2. **The two-dimensional Lorentz gas.** Replace the trees by hard disks of radius a, and allow the 'wind' particles to move in all directions on the plane. Derive an equation for $F(\mathbf{r}, \mathbf{v}, t)$. Prove an H-theorem.

3. **The diffusion coefficient.** Complete the calculation of the tagged particle diffusion coefficient. Use the Chapman–Enskog method to compute the diffusion coefficient for the moving particle in the two-dimensional Lorentz gas, and for the moving particle in the wind-tree model, assuming in each case that the scatterers are placed at random in the plane, without overlapping, and at low densities.

3

Liouville's equation

3.1 Derivation

An N-particle Hamiltonian system is specified at any instant of time by the coordinates and the momenta of all of the particles in the system. These $2dN$ variables, where d is the number of dimensions for each position or momentum vector, define a phase-space, Γ-space, and one point in this space determines all the coordinates and momenta of all of the particles. In the course of time, such a point follows a trajectory in Γ-space. A trajectory is uniquely determined by any of its points and the Hamiltonian, $H(\mathbf{r}, \mathbf{p})$:

$$\dot{\mathbf{r}}_i = \frac{\partial H}{\partial \mathbf{p}_i}, \quad \dot{\mathbf{p}}_i = -\frac{\partial H}{\partial \mathbf{r}_i}. \tag{3.1}$$

Two distinct trajectories cannot intersect or be tangent to each other, but single trajectories can be closed curves in phase-space.

An ensemble of N-particle systems is, at any instant, described as a large collection of points in Γ-space. A continuous set of points can also be taken to form an infinite ensemble. Then $\rho(\Gamma)d\Gamma$ is the probability of finding a member of the ensemble in $d\Gamma = d\mathbf{r}_1 d\mathbf{p}_1 \ldots d\mathbf{r}_N d\mathbf{p}_N$ about the point Γ. The distribution function, $\rho(\Gamma)$, changes with time as the members of the ensemble follow their individual trajectories. Time evolution neither creates nor destroys members of the ensemble, so for a region R in Γ-space, there must be a conservation law:

$$\int_R d\Gamma \frac{\partial \rho(\Gamma, t)}{\partial t} = \mathrm{Flux}|_{\mathrm{in}} - \mathrm{Flux}|_{\mathrm{out}} = \int_{\delta R} dS(\mathbf{v}\rho), \tag{3.2}$$

where $\mathbf{v} = (\dot{\mathbf{r}}_1, \dot{\mathbf{p}}_1 \ldots \dot{\mathbf{r}}_N, \dot{\mathbf{p}}_N)$, dS denotes a surface element on the boundary, δR, of the region R. In differential form, we have

$$\frac{\partial \rho}{\partial t} + \nabla_{6N} \cdot (\mathbf{v}\rho) = 0, \tag{3.3}$$

where $\nabla_{6N} = \left(\frac{\partial}{\partial \mathbf{r}_1}, \frac{\partial}{\partial \mathbf{p}_1} \cdots, \frac{\partial}{\partial \mathbf{r}_N}, \frac{\partial}{\partial \mathbf{p}_N}\right)$. This leads to a conservation equation of the form

$$\frac{\partial \rho}{\partial t} + \sum_i^N \frac{\partial}{\partial \mathbf{r}_i} \cdot (\dot{\mathbf{r}}_i \, \rho) + \sum_i^N \frac{\partial}{\partial \mathbf{p}_i} \cdot (\dot{\mathbf{p}}_i \, \rho) = 0. \tag{3.4}$$

One can readily verify that the terms on the right-hand side which don't contain derivatives of ρ cancel because of the Hamiltonian equations of motion. That is,

$$\frac{\partial}{\partial \mathbf{r}_i} \cdot \left(\frac{d\mathbf{r}_i}{dt}\right) + \frac{\partial}{\partial \mathbf{p}_i} \cdot \left(\frac{d\mathbf{p}_i}{dt}\right) = 0.$$

Then

$$\frac{\partial \rho}{\partial t} + \sum_i^N \left(\dot{\mathbf{r}}_i \cdot \frac{\partial}{\partial \mathbf{r}_i} + \dot{\mathbf{p}}_i \cdot \frac{\partial}{\partial \mathbf{p}_i}\right) \rho = 0.$$

This is *Liouville's equation*

$$\frac{d\rho(\Gamma, t)}{dt} = 0. \tag{3.5}$$

It expresses the conservation of density of members of the ensemble in Γ-space. If R in (3.2) denotes the whole Γ-space and if $\rho(\Gamma)$ falls off sufficiently rapidly, then

$$\int_{\mathrm{R}} d\Gamma \frac{\partial \rho(\Gamma, t)}{\partial t} = \frac{d}{dt} \int_{\mathrm{R}} d\Gamma \rho(\Gamma, t) = 0,$$

i.e., the total probability of finding a system somewhere in phase-space remains constant (and equal to unity if the probability density is normalized).

Equation (3.5) has many important consequences. For example, consider at $t = 0$ a set A in Γ-space and take for the initial ensemble density the characteristic function of A

$$\rho(\Gamma, 0) = \chi_A(\Gamma) = \begin{cases} 1 \text{ if } \Gamma \in A \\ 0 \text{ if } \Gamma \notin A. \end{cases}$$

The volume of A is given by $V(A) = \int_{\text{all space}} d\Gamma \chi_A(\Gamma)$. Now Liouville's equation implies $\rho(\Gamma(t), t) = \rho(\Gamma, 0)$. After some time t,

$$\rho(\Gamma, t) = \begin{cases} 1 \text{ if } \Gamma \text{ evolved from } A, \text{ or } \Gamma(-t) \in A \\ 0 \text{ if } \Gamma \text{ did not evolve from } A, \text{ or } \Gamma(-t) \notin A. \end{cases}$$

So $\rho(\Gamma, t) = \chi_{A(t)}(\Gamma)$. An integration over Γ-space gives

$$\int d\Gamma \rho(\Gamma, t) = V(A(t)). \tag{3.6}$$

From (3.5), it then follows that $dV(A(t))/dt = 0$. In other words, time evolution in the phase-space, Γ, is a volume-preserving transformation. In our subsequent discussions we will often refer to the volume of a set in Γ-space as the Liouville measure in this phase-space, and we have just proved that the Liouville measure is preserved under time evolution.

A very similar discussion can be given for the measure of regions on a constant-energy surface, that we will use to describe systems with a given and fixed energy. Consider points in phase-space that are distributed uniformly over a region where $E < H < E + \delta E$. Let dS be the differential surface area for the constant-energy surface with energy E, and let δz be the infinitesimal (perpendicular) distance between the two surfaces. Then the differential volume element for a region between the surface at E and the one at $E + \delta E$ is $dV = dS\delta z$. Using $\delta E = \delta z|\nabla H|$ we see that the volume element $dV = dS\delta E/|\nabla H|$ is conserved in the course of time. Since the left-hand side is constant in time, and since δE is fixed, one can define an invariant measure on the constant-energy surface by $\mu(dS) = dS/|\nabla H|$.

3.2 The BBGKY hierarchy equations

In our discussion of the Boltzmann equation, we concluded that this equation had to be understood as describing the average behavior of an ensemble of similarly prepared dilute gases, rather than the behavior of one particular member of the ensemble. If this is true, then we ought to be able to derive the Boltzmann equation from the Liouville equation by integrating the N-particle distribution function $\rho(\Gamma, t)$ over the phases of all but one of the particles. This is in fact true, and the work of Bogoliubov, Green and Cohen showed how this could be accomplished. These workers provided the foundation for a very advanced kinetic theory which has led to (i) a derivation of the Boltzmann equation, (ii) allowed for a generalization of the Boltzmann equation to higher densities, and (iii) uncovered a variety of physical phenomena which have been studied by means of computer simulations and experiments on laboratory systems. We will not discuss this here, since it would take us away from the main topic of the connections with dynamical systems theory, but interested readers should consult the references listed at the end of this chapter.

Here we will only show how this procedure is initiated by carrying out the integration of the phase-space distribution function over the phases of all but one of the particles. We begin by writing the Liouville equation, (3.5), as

$$\frac{\partial \rho(\Gamma,t)}{\partial t}+\sum_{i}^{N}\frac{\mathbf{p}_i}{m}\cdot\nabla_{\mathbf{r}_i}\rho(\Gamma,t)-\sum_{i<j}^{N}\theta_{ij}\rho(\Gamma,t)=0, \tag{3.7}$$

where we have used the explicit form of the total time derivative in (3.5), and defined the operator θ_{ij} by

$$\theta_{ij}=\theta(\mathbf{r}_i,\mathbf{p}_i,\mathbf{r}_j,\mathbf{p}_j)=\frac{\partial\varphi(r_{ij})}{\partial\mathbf{r}_i}\cdot\left(\frac{\partial}{\partial\mathbf{p}_i}-\frac{\partial}{\partial\mathbf{p}_j}\right),$$

and $\varphi(r_{ij})$ is the intermolecular pair potential energy, which is a function of $r_{ij}=|\mathbf{r}_i-\mathbf{r}_j|$. From our discussion of the Boltzmann equation, we know that the quantity of interest to us is $F(\mathbf{r},\mathbf{v},t)$, which we now think of as the ensemble average of the density of particles in the region $\delta\mathbf{r}\delta\mathbf{v}$ at time t. Since Γ-space is more properly formulated in terms of $\mathbf{r}$ and $\mathbf{p}$ instead of $\mathbf{r}$ and $\mathbf{v}$, we simply define a new function of position and momentum, $\hat{F}(\mathbf{r},\mathbf{p},t)$, such that

$$\hat{F}(\mathbf{r},\mathbf{p},t)\delta\mathbf{r}\delta\mathbf{p}=F(\mathbf{r},\mathbf{v},t)\delta\mathbf{r}\delta\mathbf{v},$$

and continue the discussion for $\hat{F}$. Let us now suppose that the phase-space distribution function is normalized to unity, that is,

$$\int d\Gamma\rho(\Gamma,t)=1,$$

where $d\Gamma$ is the volume element in phase-space. Then the single-particle distribution function $\hat{F}(\mathbf{r},\mathbf{p},t)$ is defined by taking the average value of the microscopic density of particles at $(\mathbf{r},\mathbf{p})$ as

$$\hat{F}(\mathbf{r},\mathbf{p},t)=\int d\Gamma\rho(\Gamma,t)\sum_{i}^{N}\delta(\mathbf{r}_i-\mathbf{r})\,\delta(\mathbf{p}_i-\mathbf{p}). \tag{3.8}$$

This then is the proper formulation of the single-particle distribution function as a phase-space average. We should expect this function to satisfy the Boltzmann equation if our system is a dilute gas. An equation of motion for $\hat{F}$ can be found by differentiating (3.8) with respect to time and by using the Liouville equation,

(3.7), to determine the time derivative of $\rho(\Gamma, t)$. When this is carried out, one obtains (the proof is left to the reader)

$$\frac{\partial \hat{F}(\mathbf{r},\mathbf{p},t)}{\partial t} + \frac{\mathbf{p}}{m} \cdot \nabla_{\mathbf{r}} \hat{F}(\mathbf{r},\mathbf{p},t) = \int d\mathbf{r}' \int d\mathbf{p}' \theta\left(\mathbf{r},\mathbf{p},\mathbf{r}',\mathbf{p}'\right) \hat{F}_2\left(\mathbf{r},\mathbf{p},\mathbf{r}',\mathbf{p}',t\right), \quad (3.9)$$

where the θ-operator is defined above, and $\hat{F}_2$ is the two-particle distribution function defined by

$$\hat{F}_2(\mathbf{r},\mathbf{p},\mathbf{r}',\mathbf{p}',t) = \int d\Gamma \rho(\Gamma,t) \sum_{i \neq j}^{N} \delta\left(\mathbf{r}_i - \mathbf{r}\right) \delta\left(\mathbf{p}_i - \mathbf{p}\right) \times \delta\left(\mathbf{r}_j - \mathbf{r}'\right) \delta\left(\mathbf{p}_j - \mathbf{p}'\right), \quad (3.10)$$

which is the average probability density of finding *two* particles in the fluid, one at $(\mathbf{r},\mathbf{p})$, and the other at $(\mathbf{r}',\mathbf{p}')$. Equation (3.9) is called the first BBGKY hierarchy equation. Hierarchy is used to denote the fact that if one wants to find the equation of motion for $\hat{F}_2$ one finds that it depends on a three-particle probability density, and so on. There is in fact an open-ended chain of equations for the $\hat{F}_n$ functions derived by Born, Bogoliubov, Green, Kirkwood, and Yvon. The extraction of the Boltzmann equation from this chain of equations is no small task, but it can be done, and a lot more besides.

We will not pursue this issue here, other than to say that the derivation of the Boltzmann equation from the hierarchy equations requires some assumptions about the properties of the n-particle distribution functions $\hat{F}_n$ at some *initial* time. These assumptions usually are that the initial values of the $\hat{F}_n$ are smooth functions of their arguments and that for initial configurations of particles where some are far from the others, then the $\hat{F}_n$ can be written as a product of lower-order distribution functions involving the separated groups of particles.

We will return to the BBGKY equations again in the context of Lorentz lattices gases in Chapter 16, and in the general discussion of the foundations of the Boltzmann equation in Chapter 17.

3.3 Poincaré recurrence theorem

One of the most interesting consequences of Liouville's theorem is the famous Poincaré recurrence theorem, which states, roughly, that almost every initial state of a system will occur again in the course of the time evolution of the system. This theorem was stated by Nietzsche who called it the *Law of Universal Return.*† We turn to a careful statement and proof.

Consider a classical system confined to a phase-space of finite measure, $\mu = \int dS/|\nabla H| < \infty$. For example, think of a system with finite energy and extension, so that for all particles i

$$0 < \frac{\mathbf{p}_i^2}{2m} < M_p,$$

and

$$0 < |\mathbf{r}_i| < M_r,$$

where M_p and M_r are finite, positive quantities. In this case, Poincaré's recurrence theorem states that *almost every point in phase-space is recurrent* (i.e., the set of points that do not return arbitrarily close to their starting point after a sufficiently long time has measure zero).

To prove it, consider a set of points, A, in phase-space with finite measure. Next, consider in A a set of points, B_0, that leave A after a time τ and that never return to A. We let the time τ be arbitrary, and B_0 is a non-recurrent set of points. We wish to show that $\mu(B_0) = 0$.

Let B_n denote the set of points that B_0 evolves to after time $n\tau$. We claim that the intersection between B_n and B_k is empty for every set of integers (k, n) with $k > n$. If this were not the case then some $x_n \in B_n \cap B_k$ would have evolved from $x_{n-1} \in B_{n-1} \cap B_{k-1}$. So, $B_{n-1} \cap B_{k-1} \neq \emptyset$. Repeating this n times we get $B_0 \cap B_{k-n} \neq \emptyset$, implying that a point in B_0 has returned to A after a time $(k-n)\tau$. This contradicts the assumption and proves the claim. Because time evolution is measure-preserving according to Liouville's equation, the total measure of the sets $B_0, \ldots, B_n$ after a time $n\tau$ is $n\mu(B_0)$. For a non-zero $\mu(B_0)$ and some large enough n, this number will exceed the total measure of Γ-space.

† In this form, even Woody Allen referred to it in one of his movies, *Hannah and Her Sisters.*

Table 3.1 *The recurrence time, t_r, for one percent density fluctuations in a spherical volume of radius a in air at standard temperature and pressure.*

a [cm]	t_r [sec]
$\sim 10^{-5}$	$\approx 10^{-11}$
$\sim 2.5 \times 10^{-5}$	≈ 1
$\sim 3 \times 10^{-5}$	$\approx 10^{6}$
$\sim 5 \times 10^{-5}$	$\approx 10^{68}$
~ 1	$\approx 10^{10^{14}}$

Therefore, considering that we have proved that no overlap can occur between the B_ks, it follows that $\mu(B_0)$ must be zero.

Poincaré's recurrence theorem has strange implications; for example, the particles in a bottle of gas that have an exceptional initial configuration will, even after equilibrium has been established, eventually return to their original positions. Very large fluctuations will occur after enough time. Zermelo used the theorem to argue that Boltzmann should not try to reconcile the second law of thermodynamics and classical mechanics. In his reply, Boltzmann compared Zermelo to a dice player who complains that a die is false because he has never thrown 1000 ones in a row, even though this possibility has non-zero measure. The recurrence of an ordinary system to an unlikely state has never been recorded, because the recurrence times are *very* large. It is still true that when a system is not in equilibrium, the most likely movement will be towards equilibrium. Poincaré's theorem just doesn't exclude the possibility that a system can also move away from it as it returns, eventually, to the vicinity of its initial state. A similar theorem can also be proved for quantum systems (see Exercise 3.2.).

Chandrasekhar has given an instructive calculation of Poincaré recurrence times. Chandrasekhar considers a spherical volume of air of radius a which is a part of a larger container of air at STP. He then computes the recurrence time, t_r, for a one percent fluctuation in the density of the air in this spherical volume about its average value. The results are given in Table 3.1.

3.4 Further reading

A reasonable derivation of Liouville's equation can be found in almost any text on statistical mechanics, for example, the two volumes of Toda *et al* [TKS92]. The BBGKY hierarchy equations and the Bogoliubov approach to kinetic theory are described in Bogoliubov's original article [Bog62], as well as in the more recent papers of Dorfman and van Beijeren [DvB77], Cohen [Coh93, Coh97a], Ernst [Ern98], and others.

Most books on ergodic theory contain a proof of the Poincaré recurrence theorem. The book by Katok and Hasselblatt [KH95] is a good source. Brush [Bru76] gives a nice history of this theorem and connects it to the work of Nietzsche, and the readers may wish to see Woody Allen's version for themselves. The estimates of simple recurrence times are given in the article by Chandrasekhar in the collection of Wax [Wax54]. The Ehrenfest urn model (Exercise 3.1.) is described in Kac's book [Kac76], and in the paper of Kac in the collection of Wax [Wax54].

3.5 Exercises

1. **The Ehrenfest urn model.** Here is an experiment simple to simulate using a computer. Consider two urns, I and II, and a bag. N balls, labeled $1, \ldots, N$, are distributed between the urns. The bag contains N pieces of paper that carry numbers $1, \ldots, N$. At each time step, somebody draws a number out of the bag at random, moves the corresponding ball to the other urn, and then puts the piece of paper back in the bag. This is repeated a large number of times. Consider the (absolute) difference $|N_{\mathrm{I}}(t) - N_{\mathrm{II}}(t)|$ between the number of balls in I and in II as a function of time. How would you characterize the behavior for short and long times? How can you relate the behavior of the distribution of the balls in the containers for this model to a resolution of the Zermelo and Loschmidt paradoxes in the theory of the Boltzmann equation? Notice that the chance that a ball will move from the fuller urn to the emptier urn is always greater than the chance that the opposite will happen, therefore the system has a tendency to approach an equilibrium, where the balls are distributed equally between the urns. Fluctuations will also occur, even very large ones.

Analytically, it is possible to formulate this problem as a Markov process to obtain a description of the process. This is described in the article by Kac in [Wax54].

2. **Poincaré recurrence theorem for quantum systems.** Consider a Schroedinger wave function, $\Psi(\mathbf{r}, t)$. Suppose that $\{\phi_n(\mathbf{r})\}$ is a complete set of states, and that the expansion of Ψ in terms of this set,

$$\Psi(\mathbf{r}, t) = \sum_n a_n e^{i\omega_n t} \phi_n(\mathbf{r}),$$

converges uniformly at all times. Suppose also that

$$\int |\Psi(\mathbf{r}, t)|^2 d\mathbf{r} = 1, \quad \text{and} \quad \int |\phi_n|^2 d\mathbf{r} = 1.$$

Then show that $\Psi(\mathbf{r}, t)$ is an almost periodic function of time.

4
Boltzmann's ergodic hypothesis

4.1 Introduction

In Chapter 1, we described the approach of Maxwell and Boltzmann in their attempt to justify the use of statistical mechanics on the basis of the dynamical and statistical properties of mechanical systems composed of large numbers of particles. One feature of this analysis is the *ergodic hypothesis*. That is, that starting from almost all initial points, the phase-space trajectory will explore many regions of phase-space over the time-scales of a laboratory experiment. Then, if the system spends an amount of time in a region that is proportional to some natural probability measure, it will spend most of its time in regions where the macroscopic variables take their equilibrium values, since those regions occupy most of the phase-space. Here, we develop this idea further using some new mathematical notions. Instead of looking at the motion of systems over finite times, we will be considering infinite time averages. This is an issue that requires further exploration, since laboratory experiments take only finite times, of course. However, it is not unreasonable to expect that the infinite-time limit really means *a time long compared to some characteristic microscopic time*, and that the laboratory experiments take place on even longer time-scales, where the underlying ergodic behavior of the system can make itself manifest. Of course, it is necessary to see if this expectation is actually met in a real system, and to examine the consequences if it is not. It is well known that the application of mathematical ideas to statistical mechanics requires the application of some 'tact'†, and with this in mind, we tactfully

† One of the favorite expressions of my thesis advisor, the late Theodore Berlin, was 'Statistical mechanics is the science of tact'.

take infinite-time limits in what follows.†

4.2 Equal times in regions of equal measure

Mechanical systems – isolated ones, at least – are time reversible and recurrent. However, our observations are that large isolated systems often reach a state of thermodynamic equilibrium. How do we explain our observations in such a way that there are no contradictions with the laws of mechanics? Boltzmann's proposed resolution adopts the following lines:

1. Equilibrium statistical mechanics can be formulated in terms of microcanonical ensemble averages, by using the invariant measure $d\mu = \mu(dS) = dS/|\nabla H|$. The ensemble average of a phase-space function, $F(\Gamma)$, is

$$\langle F(\Gamma)\rangle_{\text{m.c.}} = \frac{\int_{H=E} d\mu F(\Gamma)}{\int_{H=E} d\mu}. \tag{4.1}$$

The number of states with energy E is

$$\Omega(E) = \int_{H=E} d\mu, \tag{4.2}$$

and the thermodynamic entropy, S, is given by $S = k_{\text{B}} \ln \Omega(E)$.

As an aside, we mention that one can formulate the microcanonical ensemble in terms of a phase-space density $\rho(\Gamma) = \delta(H - E)$. Then

$$\langle F(\Gamma)\rangle_{\text{m.c.}} = \frac{\int d\Gamma F(\Gamma)\delta(H-E)}{\int d\Gamma \delta(H-E)}. \tag{4.3}$$

2. If in the laboratory one were to make very precise measurements of some quantity, $F(\Gamma_t)$, where Γ_t is the location of the phase point of the system at time t, e.g., the force per unit area at time t on a piston, the values would show wild fluctuations about a more slowly varying 'mean' quantity as molecules collide with the piston. However, we might identify the thermodynamic

† We remind the reader that we are typically not interested in the trajectory of a system in the full phase-space, but rather on the projection of the trajectory onto some subspace of relevant variables. We would hope that this 'projected' trajectory is ergodic on some reasonable time-scale. An example of this 'hoped for' behavior will be discussed in Sec. 17.1.

value of some property for one system with the time average $\overline{F}(\Gamma)$ defined by

$$\overline{F}(\Gamma) = \lim_{T\to\infty} \frac{1}{T}\int_0^T F(\Gamma_t)dt, \tag{4.4}$$

where, as mentioned above, the infinite-time limit should be taken somewhat tactfully.

Boltzmann realized that he could identify the microcanonical ensemble average, (4.1), with the *infinite-time* average, (4.4), that is

$$\overline{F}(\Gamma) = \langle F(\Gamma)\rangle_{\text{m.c.}}, \tag{4.5}$$

if he made the *hypothesis* that the trajectory of a typical point on the constant-energy surface (that is, all points except for a set of points of zero measure) spends equal time in regions of equal measure. This hypothesis was called the *ergodic hypothesis* by Boltzmann and it is of central interest for the foundations of statistical mechanics.

To see how this hypothesis works, subdivide the constant-energy surface into a fine grid. The average of $F(\Gamma)$ in each grid region i is F_i (see Fig. 4.1). Then

$$\frac{1}{T}\int_0^T F(\Gamma_t)dt \approx \sum_i \frac{\tau_i}{T}F_i, \tag{4.6}$$

where τ_i/T is the fraction of the time that the trajectories spend in region i between $t = 0$ and T. Using the ergodic hypothesis, one can write

$$\frac{\tau_i}{T} = \frac{\mu_i}{\mu(\mathcal{E})}, \tag{4.7}$$

where $\mathcal{E}$ represents the surface with constant energy E. Consequently,

$$\overline{F}(\Gamma) = \sum_i \frac{\mu_i}{\mu(\mathcal{E})}F_i = \langle F(\Gamma)\rangle_{\text{m.c.}}. \tag{4.8}$$

Thus, equilibrium statistical mechanics could be justified, for isolated systems, if one could prove that the ergodic hypothesis is correct for a large class of physical systems. As we mentioned in Chapter 1, this may indeed be asking too much of the dynamics of systems of physical interest. That is, it may be that some

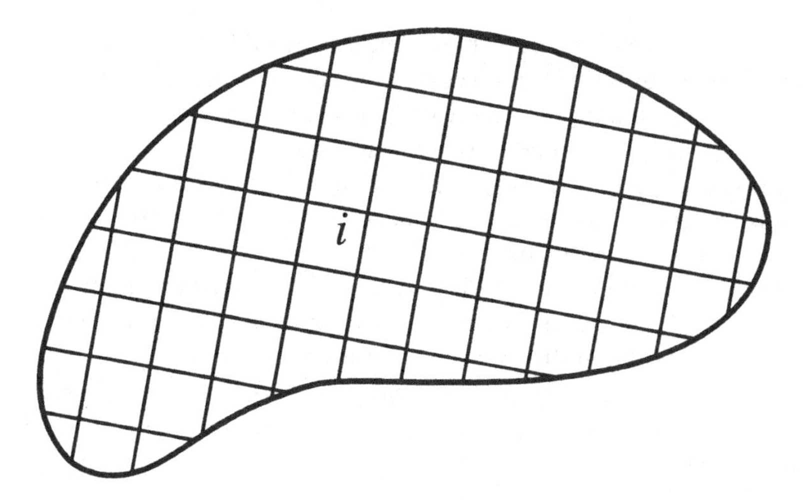

Fig. 4.1 The division of the constant energy surface into small regions.

projection of the system's trajectory on a phase-space of lower dimensions behaves ergodically in such a lower-dimensional space, and this may be enough for our purposes. In such a case, the ergodic analysis of Boltzmann would still apply in the appropriate lower-dimensional space.

One should also not ignore the fact that ergodic or ergodic-like behavior may not always be needed to justify the use of statistical methods. For example, no system of physical interest is really isolated from the rest of the universe. Thus it may be that the ergodic approach to developing the foundations for statistical mechanics is irrelevant and/or unnecessary, and the approach to equilibrium is governed, or at least produced by, the random perturbations on the systems caused by its environment or some other source. Our approach here is to explore the consequences of the chaotic behavior of many deterministic mechanical systems. If, in fact, one could prove the truth of the ergodic hypothesis for systems of physical interest, and make the appropriate arguments about time-scales, then the mechanical properties of the system and the fact that the system is composed of a large number of degrees of freedom could be held responsible for the existence of a thermodynamic equilibrium state. Then one would not need a *deus ex machina*, and the results of external perturbations would not be crucial for our understanding of the success of statistical thermodynamics. If, on the other hand, one could not show that the system explores phase-space on reasonable time-scales and spends equal times in regions of equal measures, either in the full phase-space or in an appropriate lower-dimensional space, for any interesting system, then external sources of randomness would be absolutely necessary and we would have to look to them for a serious justification of our methods. In any case, random external perturbations can help make non-ergodic systems look ergodic.

Having now stated mathematically what one needs to prove as an ergodic theorem, we now turn to seeing what one can do in this direction.

4.3 The individual ergodic theorem

In 1931, Birkhoff proved a very important theorem which allows us to make some of the ideas of Boltzmann more precise. Although Birkhoff's theorem is still a long way from what is needed in Boltzmann's picture, it does define in a useful way the dynamical properties that a system must possess in order to be ergodic in the sense of Boltzmann.

This theorem is concerned with the properties of individual trajectories in Γ-space. Suppose we are considering a mechanical system with an invariant measure, and suppose that we consider some dynamical function $F(\Gamma)$ defined on the constant-energy surface and satisfying the condition

$$\int d\mu |F(\Gamma)| < \infty.$$

Birkhoff's theorem then states that

(Birkhoff's individual ergodic theorem) *The long-time average* $\overline{F}(\Gamma) = \lim_{T\to\infty} \frac{1}{T}\int_0^T dt F(\Gamma_t)$ *exists almost everywhere on the constant-energy surface, that is, for almost every starting point* Γ.

$\overline{F}(\Gamma)$ may depend on the trajectory, but not on the initial point of the trajectory. This can be easily seen because the time average of a function starting at phase-space point Γ_{t_0} can be written as

$$\begin{aligned}\overline{F}(\Gamma_{t_0}) &= \frac{1}{T}\int_0^T dtF(\Gamma_{t_0+t}) = \frac{1}{T}\int_{t_0}^{T+t_0} dtF(\Gamma_t) \\ &= \frac{1}{T}\left[\int_{t_0}^{0} F(\Gamma_t)dt + \int_0^T F(\Gamma_t)dt + \int_T^{T+t_0} F(\Gamma_t)dt\right].\end{aligned}$$

In the limit that $T \to \infty$ the first and the third terms tend to zero, the second to $\overline{F}(\Gamma)$. So the long-time average is constant on a trajectory $\overline{F}(\Gamma) = \overline{F}(\Gamma_{t_0})$, and *one can interchange the average over the invariant measure, that is, the microcanonical average of $F(\Gamma)$, with the time average,*

$$\int d\mu F(\Gamma) = \int d\mu \overline{F}(\Gamma).$$

Finally, *the constant-energy surface can be replaced by any 'invariant' set of finite measure, and the microcanonical average is replaced by an average over the invariant set, using the measure μ.*

By an invariant set we mean a region in phase-space such that all points in the set remain in the set during the course of their time evolution, and the measure of this set of points does not change with time. This final point will be important for us when we consider some simple cases, such as the baker's transformation, where it is possible to discuss ergodic behavior in some detail.

Birkhoff's theorem is not sufficient to show that systems of physical interest are ergodic, in the sense of Boltzmann, because the time average still depends on the trajectory and is not necessarily equal to the ensemble average. However, this gives us the opportunity to *define* a system that satisfies Boltzmann's ergodic hypothesis. We say that a system is *ergodic* if the long-time average of a function, $\overline{F}(\Gamma)$, is a constant which we denote by $\tilde{F}$ on the constant-energy surface.

An ergodic system is such that the time average of a dynamical quantity is equal to its microcanonical average, for one easily sees that $\int d\mu F(\Gamma) = \int d\mu \overline{F}(\Gamma) = \tilde{F} \int d\mu$. So,

$$\tilde{F} = \frac{\int d\mu F(\Gamma)}{\int d\mu} = \langle F(\Gamma)\rangle_{\text{m.c.}}.$$

Moreover, Boltzmann's hypothesis that the time spent in a region of the constant-energy surface is proportional to its microcanonical measure is indeed true for an ergodic system. To prove this, refer to the grid picture of the last section (Fig. 4.1), and take for $F(\Gamma)$ the characteristic function of a grid region i, $\chi_i(\Gamma)$, where

$$\chi_i(\Gamma) = \begin{cases} 1 & \text{if } \Gamma \in \text{region } i \\ 0 & \text{if } \Gamma \notin \text{region } i \end{cases} \tag{4.9}$$

Then $\bar{\chi}(\Gamma) = \lim_{T\to\infty} \frac{1}{T} \int_0^T dt \chi_i(\Gamma_t) = \lim_{T\to\infty} \tau_i/T$, which is the fraction of time that the system spends in region i, and also $\bar{\chi}(\Gamma) = \langle\chi(\Gamma)\rangle_{\text{m.c.}} = \int d\mu \chi_i(\Gamma)/\int d\mu = \mu_i/\mu(\mathcal{E})$.

As an example of an ergodic system consider a circle with circumference 1. The time is a discrete variable. In one time step, the points $x \in [0,1]$ on the circle are rotated as

$$\begin{aligned} \phi: &\quad x \to \phi(x) = x + \alpha \quad (\text{mod } 1), \\ \phi^2: &\quad x \to x + 2\alpha \quad (\text{mod } 1), \end{aligned}$$

and so on (let us call this the *rotation map*). If α is a rational number, n/m, where n and m are integers, then after a finite number of steps, m, ϕ^m maps all points to their initial positions. A trajectory starting at some initial point will be periodic and will not spend equal times in regions of equal (Lebesgue) measures, and hence the trajectory is not in any way dense on the circle. In this case the system is not ergodic. But if α is irrational, no trajectory is periodic and the system is ergodic! To see this, consider the time average of an integrable function on a trajectory starting at x,

$$\overline{F}(x) = \lim_{N\to\infty} \frac{1}{N} \sum_{g=0}^{N-1} F(\phi^g(x)) = \lim_{N\to\infty} \frac{1}{N} \sum_{g=0}^{N-1} F(x + g\alpha). \tag{4.10}$$

Consider the Fourier series for $F(x)$, $F(x) = \sum_{n=-\infty}^{\infty} a_n e^{2\pi i n x}$, so that

$$F(x + g\alpha) = \sum_{n=-\infty}^{\infty} a_n e^{2\pi i n x + 2\pi i n \alpha g}. \tag{4.11}$$

By substituting this Fourier sum in the expression for the time average, we obtain

$$\begin{aligned}\overline{F}(x) &= \lim_{N\to\infty}\frac{1}{N}\sum_{g=0}^{N-1}\sum_{n=-\infty}^{\infty} a_n e^{2\pi i n x + 2\pi i n\alpha g} \\ &= a_0 + \lim_{N\to\infty}\frac{1}{N}\sum_{n\neq 0} a_n \frac{1-e^{2\pi i\alpha N n}}{1-e^{2\pi i\alpha n}} e^{2\pi i n x}. \qquad (4.12)\end{aligned}$$

The terms with $|n| \geq 1$ don't survive the limit $N \to \infty$. Thus, $\overline{F}(x) = a_0$ for α irrational. In this case the time average is a constant – independent of the starting point x – and the system is ergodic, where the ensemble average is the average over the circle with the usual Lebesgue measure,

$$\overline{F}(x) = \int_0^1 dx F(x) = \langle F(x)\rangle_{\text{m.c.}}. \qquad (4.13)$$

Sequences of numbers may also have ergodic properties (see Exercise 4.1..

We now can fill in some details of Boltzmann's picture: If we can prove that a system is *ergodic* then we know that its time average properties are equal to the appropriate microcanonical ensemble average properties, and equilibrium statistical mechanics can be used to determine the time average properties of the system. Boltzmann then argued, as we know from all of our statistical mechanics courses, that if (a) the system spends equal times in regions of equal measures, and (b) the region of the energy surface that corresponds to a macroscopic equilibrium state i.e., with Maxwell–Boltzmann velocity distribution and equilibrium many-particle distribution functions) has overwhelmingly the largest measure, then all of the usual results of equilibrium thermodynamics follow immediately, provided the macroscopic variables take statistically sharp values. The Poincaré recurrence issue is not a problem since it is of no consequence if a phase point returns to its initial region every once in a while, and the reversibility of the dynamics is no problem either, for the equilibrium state at least, since in this picture both the forward and time-reversed motion will eventually have the same – equilibrium – values for time averages.†

† We have not yet proved this, but it is an easy consequence of ergodicity and we will prove it when we look more carefully at how one proves that interesting systems are ergodic.

What remains to be done then is to determine whether or not systems of physical interest are ergodic, and if so, in what sense and on what time-scales. Before we turn to this problem, it is appropriate to consider a somewhat deeper idea of mixing, and also to discuss an equivalent way of characterizing ergodic systems as being metrically indecomposable. This we shall do in the next chapter.

4.4 Further reading

A nice discussion of the role of ergodicity in statistical mechanics is given by Ruelle [Rue91], and a more technical discussion, by Toda *et al.* [TKS92]. More mathematical approaches can be found in Katok and Hasselblatt [KH95], and by Pollicott and Yuri [PY98], as well as in the books by Halmos [Hal56], and by Arnold and Avez [AA68].

4.5 Exercises

1. **Ergodic properties of sequences of numbers.**† Consider all integral powers of 2 : $2^0, 2^1, 2^2, 2^3, \dots$. Write down the first digit of each power in a sequence:

$$1, 2, 4, 8, 1, 3, 6, 1, 2, 5, 1, \dots .$$

Calculate the frequency at which a '7' appears in this sequence. *Hint:* This is a consequence of the ergodic theorem, and you need to consider that an integer k appears whenever

$$k \cdot 10^r < 2^m < (k+1) \cdot 10^r$$

for some integer r.

† This problem is taken from the book of Arnold [Arn89], and used with the permission of Springer-Verlag.

5
Gibbs' picture: mixing systems

5.1 The definition of a mixing system

While Boltzmann fixed his attention on the motion of the phase point for a single system and was led to the concept of ergodicity, Gibbs took another approach to the same problem. Since one never knows precisely what the initial phase point of a system is, Gibbs decided to consider the average behavior of a set of points on the constant-energy surface with more or less the same macroscopic initial state. Without worrying too much about how such a set might be defined precisely, let's consider an initial set of points A. As the set travels through Γ-space, it changes shape but its measure stays the same, $\mu(A) = \mu(A_t)$. The set gets stretched and folded and may eventually appear on a coarse enough scale to fill the energy surface uniformly. However, the set A_t has the same topological structure as the set A and the initial set is not 'forgotten', in the sense that a time-reversal operation on the set A_t will produce the set A. There is a nice lecture-demonstration apparatus that illustrates this time-reversal operation: A drop of immiscible ink is added to a container of glycerine. If you stir the glycerine slowly, the drop will stretch and form a thin line. Eventually it seems to fill the whole space, but if the stirring is reversed, the initial configuration of the drop of ink surprisingly reappears.

Gibbs thought that the apparently uniform distribution of the set A_t on the energy surface was the key to understanding how mechanically reversible systems could approach an equilibrium state. To make this idea more precise, Gibbs called a system *mixing* if the following identity holds for each set B of non-zero measure:

$$\lim_{t\to\infty} \frac{\mu(B \cap A_t)}{\mu(B)} = \frac{\mu(A)}{\mu(\mathcal{E})}, \tag{5.1}$$

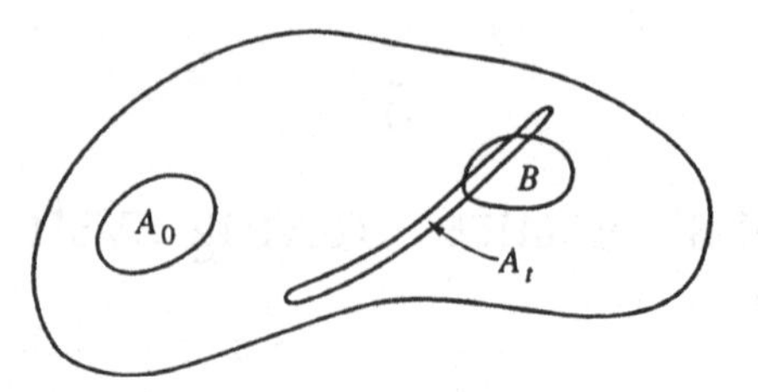

Fig. 5.1 The time development of phase-space regions and the mixing property.

where $\mathcal{E}$ represents the entire constant-energy surface, and $\mu(\mathcal{E})$, the measure of the entire surface, and is assumed to be finite (see Fig. 5.1). As we will see presently, the requirement that a system be mixing is a stronger condition than ergodicity. However, more can be said about the approach to equilibrium for a mixing system than for an ergodic one. To discuss the difference between ergodic and mixing systems we need the notion of *metric indecomposability*.

A *metrically decomposable* system is one for which there exists a subdivision of the constant-energy surface (or in general, Γ-space) into two regions of non-zero measure, each of which is invariant under the mechanical flow in Γ-space. That is, a phase point starting out in one region will always stay in that region. A system is ergodic if and only if it is metrically *in*decomposable:

1. Decomposable → non-ergodic: On a surface that can be subdivided into two invariant regions 1 and 2, let $F(\Gamma)$ be the characteristic function of region 1,

$$F(\Gamma) = \chi_1(\Gamma). \tag{5.2}$$

Due to the invariance, the time average of F is

$$\begin{aligned} \overline{F}(\Gamma) &= \lim_{T\to\infty} \frac{1}{T} \int_0^T \chi_1(\Gamma_t)dt = \chi_1(\Gamma) \\ &= \begin{cases} 0 & \text{if } \Gamma \in 2 \\ 1 & \text{if } \Gamma \in 1. \end{cases} \end{aligned} \tag{5.3}$$

Since this is not constant over Γ-space, the system is not ergodic. The converse argument is as follows.

2. Non-ergodic → decomposable: For a non-ergodic system $\overline{F}(\Gamma)$ is not constant, but depends on the trajectory. For *some* γ, the sets

of points $\{\Gamma|\overline{F}(\Gamma) \geq \gamma\}$ and $\{\Gamma|\overline{F}(\Gamma) < \gamma\}$ both contain complete trajectories, because $\bar{F}(\Gamma)$ is constant on a trajectory. Therefore, the sets are invariant and form a decomposition into two sets, each of positive measure. If this were not true for some γ then our assumption that $\bar{F}$ is not constant would be violated.

Now it will be shown that mixing implies ergodicity. Consider a mixing system and an invariant set $A = A_t$. Then, for all B, (5.1) gives

$$\lim_{t\to\infty} \frac{\mu(A_t \cap B)}{\mu(B)} = \frac{\mu(A)}{\mu(\mathcal{E})} \tag{5.4}$$

or, equivalently,

$$\lim_{t\to\infty} \mu(A_t \cap B) = \frac{\mu(A)\mu(B)}{\mu(\mathcal{E})}. \tag{5.5}$$

If we set $B = A = A_t$ (since A is an invariant set), then $A_t \cap B = A$ and

$$\mu(A) = \frac{\mu(A)\mu(A)}{\mu(\mathcal{E})}. \tag{5.6}$$

This equation has two solutions:

1. $\mu(A) = 0$: then one trivial invariant set is a set of zero measure, and the other solution is
2. $\mu(A) = \mu(\mathcal{E})$: the invariant set is effectively the whole energy surface.

Therefore if a system is mixing, the only invariant set with positive measure is the constant-energy surface. Any other invariant set must have zero measure. Such sets might be a countable set of periodic orbits. Consequently, a mixing system is ergodic, but the converse is not true, as can be seen from a simple counterexample. Consider the rotations on the circle discussed in the previous chapter. When an initial point is moved an irrational distance around the circle the transformation is ergodic, and the point in the course of its trajectory spends an equal amount of time in all regions of the circle with the same measure. However, to determine if this system is mixing, we need to look at the time evolution of a set of points. Suppose we take a small connected interval on the circle as our initial set A. Then the set A_t will remain a small connected set of points rotating rigidly around the circle. As a result of this, it is easy to see that the limit as $t \to \infty$ of the measure of the set $A_t \cap B$ does not exist for any arbitrary set B of positive measure on the circle. Hence, the irrational rotations on the circle are ergodic but not mixing.

5.2 Distribution functions for mixing systems

Consider an isolated system that initially is in a nonequilibrium state. For example, it may have a temperature- or pressure gradient. We wish to calculate the ensemble average at time t of a dynamical quantity $F(\Gamma)$,

$$\langle F(\Gamma)\rangle_t = \frac{\int d\mu \rho(\Gamma,t)F(\Gamma)}{\int d\mu \rho(\Gamma,t)}, \tag{5.7}$$

where $\rho\,(\Gamma,t)$ is the time-dependent phase-space distribution function for the system. It satisfies Liouville's equation,

$$\frac{d}{dt}\rho(\Gamma,t) = 0. \tag{5.8}$$

Suppose the system is mixing. Again, think of the constant-energy surface as a patchwork, where for each patch, i, there is a characteristic function, χ_i. Then $F(\Gamma)$ can be described approximately by typical values of F on each patch, F_i:

$$F(\Gamma) \approx \sum_j F_j \chi_j(\Gamma). \tag{5.9}$$

We can express $\rho(\Gamma,0)$ in this way also as

$$\rho(\Gamma,0) = \sum_j \rho_j \chi_j(\Gamma). \tag{5.10}$$

Now, from Liouville's equation, (3.5), it follows that $\rho(\Gamma,t) = \rho(\Gamma_{-t},0)$, where Γ_{-t} is the phase point that evolves to Γ after a time interval t. So

$$\rho(\Gamma,t) = \sum_j \rho_j \chi_j(\Gamma_{-t}). \tag{5.11}$$

When this expression is substituted into (5.7), we find

$$\langle F(\Gamma)\rangle_t = \frac{\sum_j \sum_k F_j \rho_k \int d\mu \chi_j(\Gamma)\chi_k(\Gamma_{-t})}{\sum_k \rho_k \int d\mu \chi_k(\Gamma_{-t})}. \tag{5.12}$$

The integrand in the numerator is 1 if Γ is in the jth patch and Γ_{-t} is in the kth patch, otherwise it is zero. The second condition means that $\Gamma \in k_t$, the kth patch after a time t. The integral becomes

$$\int d\mu \chi_j(\Gamma)\chi_k(\Gamma_{-t}) = \mu(j \cap k_t). \tag{5.13}$$

Because the system is mixing,

$$\mu(j \cap k_t) \to \frac{\mu(j)\mu(k)}{\mu(\mathcal{E})} \qquad \text{as } t \to \infty. \tag{5.14}$$

The denominator doesn't actually depend on the time and can also be approximated:

$$\begin{aligned} \lim_{t\to\infty} \langle F(\Gamma)\rangle_t &= \frac{1}{\sum_k \rho_k \mu(k)} \sum_j \sum_k F_j \rho_k \left(\frac{\mu(j)\mu(k)}{\mu(\mathcal{E})}\right) \\ &= \frac{1}{\mu(\mathcal{E})} \sum_j F_j \mu(j) = \langle F(\Gamma)\rangle_{\text{m.c.}}. \end{aligned} \tag{5.15}$$

This shows that the time-dependent ensemble average of any well-behaved dynamical quantity for a mixing system approaches an equilibrium value as $t \to \infty$. This is equivalent to the statement that for a mixing system, the time-dependent phase-space distribution function approaches its equilibrium value as $t \to \infty$. Here we have a nice illustration of the value for physics of what mathematicians call a *weak limit*. That is, suppose that the initial distribution function is concentrated on some set A and zero elsewhere. Then, in the course of time, the distribution function is non-zero on some very long extended set determined by the time evolution of A and zero elsewhere, as illustrated in Fig. 5.1. This long 'stringy' set has the same measure as A itself, of course. What we have just shown, though, is that averages computed with the distribution function on this 'stringy' set are, after a long enough time, equal to averages computed with a smooth equilibrium distribution function. In this *weak* sense we can say that the distribution function approaches its equilibrium value. This phenomenon is at the core of what are called SRB (Sinai–Ruelle–Bowen) measures which we will discuss in later chapters.

5.3 Chaos

We have now reached a point in our discussions where we can begin to understand some of the central ideas of this approach to nonequilibrium statistical mechanics based on dynamical systems theory. We have argued that in order for the dynamical functions of an *isolated* system to approach well-defined equilibrium values, and for the system itself to be described by a distribution function that approaches an equilibrium distribution, the system should be mixing on laboratory time-scales, at least on some subspace of the phase-space, if not on the full phase-space. This condition on the dynamical properties has some astounding consequences which have been described, collectively, as *chaotic* behavior. To be more precise, consider some set of points on the constant-energy surface with arbitrarily small, but still positive measure. If the system is mixing then no matter how small the measure of the set of points, the set will be spread out uniformly on the energy surface as the system develops in time. This means that any two points in this set will typically head to very different regions of phase-space no matter how close the two points may have been initially. The trajectories of the points must then be very sensitive functions of the initial coordinates and momenta of the particles of the system. This sensitive dependence of trajectories on initial conditions is the characteristic feature of chaotic systems.

A simple example of this separation of nearby trajectories is illustrated in Fig. 5.2. Here we illustrate the classical Lorentz gas. A collection of fixed scatterers is placed at random in space – here we use hard disks on a plane – and a point particle moves with constant velocity along straight-line paths until it collides with a scatterer whereupon it makes a specular, energy-conserving collision. This is a well-known model for studying the diffusion of electrons in amorphous materials. We show two trajectories that start at the same point in space, but with slightly different directions of the velocity vector. Note that the two trajectories separate at each collision with a scatterer in exactly the same way that a beam of light is dispersed by a convex mirror. After a number of collisions the two trajectories are sufficiently far apart that one trajectory includes a collision with a scatterer that is just missed by the other one. After that, the two trajectories are completely different, each exploring a different region of space. Further, one can argue that the time-scale for this separation of trajectories to make itself manifest should be on the order of a few mean free times between collisions of the moving particle with the

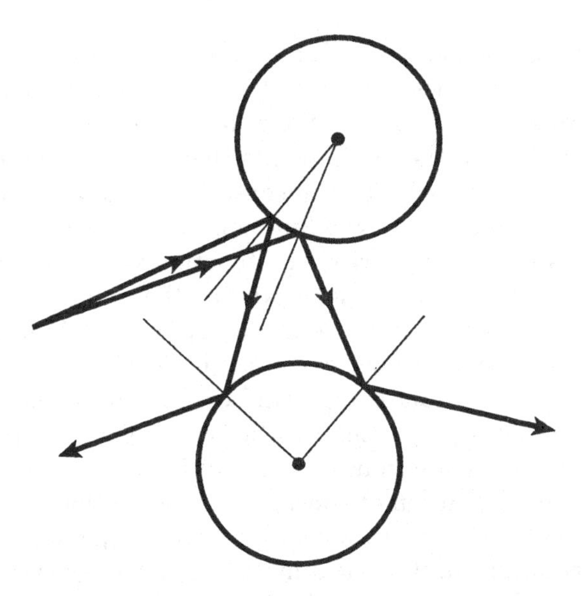

Fig. 5.2. Separation of trajectories in a Lorentz gas.

scatterers. Simple estimates show that as few as ten collisions or so are enough for the separation of very close trajectories to become on the order of macroscopic scales. Something like this takes place in the much higher-dimensional phase-space for systems where all of the particles are moving, and this picture is at the heart of the ergodic theory of billiard systems developed by Sinai and others.

Moreover, one can argue that most of the constants of the motion for a mixing system must be very singular functions of the initial coordinates and momenta, as well. One knows from elementary mechanical considerations that a system of N particles in three dimensions has $6N$ constants of the motion, determined by the initial values of all of the coordinates and momenta. We can then manipulate these constants to get a new set of constants of the motion with more interesting physical meanings. One of these new constants is the total energy of the system, another constant fixes the initial point on the trajectory. Except for perhaps a few other constants of the motion, such as the total momentum, we cannot fix the other constants of the motion when we prepare a system in the laboratory. Now if the system is ergodic, its trajectory must pass arbitrarily close to almost every point on the constant-energy surface. All of these other points may have very different values for these 'uncontrolled' constants of the motion, so the values of these other constants of the motion must be such singular functions of the phase points that in any close neighborhood of almost every point the 'uncontrolled' constants take on all possible values! In the theory of functions of complex variables one defines an 'essential singularity' of a function as a point where the function takes on all values (for example, the point $z = 0$ for the function $f(z) = \exp(-1/z)$). We might say that the 'uncontrolled' constants of the motion have essential singularities at almost every point on the constant-energy surface. In view of the fact that functions with such properties may not exist as well defined functions, or even as distributions, only analytic functions are usefully in the standard definition of constants of the motion.

Consequently, we may reasonably be able to justify the methods of nonequilibrium statistical mechanics for large systems that show sensitive dependence on initial conditions, and very complex behavior of the phase-space trajectories, such that we can prove them to be mixing systems, in some sense.

As the first example of the application of the idea that phase-space trajectories may depend very sensitively upon the parameters used to specify the state of the system, we turn in the next chapter to one of the most interesting controversies in nonequilibrium statistical mechanics: van Kampen's objections to linear response theory.

5.4 Further reading

Good discussions of mixing system systems can be found in the books of Arnold and Avez [AA68] and Katok and Hasselblatt [KH95]. Introductions to chaos theory are to be found in the books of Robinson [Rob99], Ruelle [Rue91], Ott [Ott92], Schuster [Sch89], Nicolis [Nic95], and Alligood *et al.* [ASY97]. The papers in the books edited by Sinai [Sin89, Sin91] provide the fundamental mathematical structure for chaos theory. See also the article by Lanford [Lan83]. The book by Mackey [Mac92] has very interesting discussions of ergodicity and mixing, as well as of several related topics, with a special interest in a basic understanding of the second law of thermodynamics.

6
The Green–Kubo formulae

6.1 Linear response theory

Linear response theory describes the changes that a small applied external field induces in the macroscopic properties of a system in equilibrium. The external field is supposed to be turned on at some initial time, when the system is in equilibrium, and then treated as a perturbation. As an example of linear response theory, we show how to use it to obtain the time–correlation-function expression – often called the Green–Kubo expression – for the electrical conductivity of a system that contains charged particles. If the applied electric field is small enough that heating effects can be ignored, then Ohm's law can be expressed as $\mathbf{J}_e = \sigma\mathbf{E}$, where $\mathbf{J}_e$ is the electrical current density, $\mathbf{E}$ is the applied electric field, and σ is the electrical conductivity that we wish to compute. The time–correlation formula is an example of a set of formulae which relate transport coefficients in a fluid to time integrals of time–correlation-functions. In the last section of this chapter, we will give an example of the derivation of such formulae for the case of tagged particle diffusion. First, we wish to examine one particular derivation of the formula for the electrical conductivity which has provoked a great deal of very instructive discussion, which, in turn, is closely connected to the general theme of this book.

Suppose that a small electric field is applied to a system of charged particles at $t = 0$. Liouville's equation, (3.5), can be written in the form

$$\frac{\partial \rho}{\partial t} + \{\rho, H\} = 0, \tag{6.1}$$

where $\{,\}$ is the Poisson bracket

$$\{A, B\} = \sum_i \left[\frac{\partial A}{\partial \mathbf{r}_i} \cdot \frac{\partial B}{\partial \mathbf{p}_i} - \frac{\partial A}{\partial \mathbf{p}_i} \cdot \frac{\partial B}{\partial \mathbf{r}_i}\right]. \tag{6.2}$$

The Hamiltonian, H, depends on the time and it consists of two parts $H = H_0 + H_1$, the field-free Hamiltonian with the kinetic and potential energy of the particles,

$$H_0 = \sum_i \frac{p_i{}^2}{2m} + \sum_{i<j} \phi_{ij},$$

and a term for the interaction of the particles with the electrical field,

$$H_1 = -\sum_i q_i(\mathbf{r}_i \cdot \mathbf{E}(t)),$$

where q_i is the charge of the ith particle. Since the electric field is weak, an expansion of the phase-space density ρ can be obtained as an expansion in powers of the electric field as

$$\rho = \rho_0 + \rho_1 + \cdots. \tag{6.3}$$

Substituting this in Liouville's equation and examining each power of $\mathbf{E}$, we obtain the zeroth-order equation

$$\frac{\partial \rho_0}{\partial t} + \{\rho_0, H_0\} = 0. \tag{6.4}$$

The first-order terms give

$$\frac{\partial \rho_1}{\partial t} + \{\rho_0, H_1\} + \{\rho_1, H_0\} = 0 \tag{6.5}$$

and so on. Now, the following arguments are used to obtain Ohm's law:

1. ρ_0 is the equilibrium distribution of the system without the electric field. A canonical ensemble distribution is taken to be

$$\rho_0 = \frac{1}{Z} e^{-\beta H_0},$$

where the normalization factor, Z, is the partition function.

Since the zeroth-order equation can be solved, the inhomogeneous term in the first order equation, (6.5) is known, and (6.5) can be written as

$$\frac{\partial \rho_1}{\partial t} + \mathcal{L}\rho_1 = -\{\rho_0, H_1\}. \tag{6.6}$$

Here $\mathcal{L}$ is a linear operator, called the *Liouville operator*, defined by the Poisson bracket, $\mathcal{L}\rho_1 = \{\rho_1, H_0\}$.

2. Equation (6.5) is integrated to give

$$\rho_1(t) = e^{-(t-t_0)\mathcal{L}}\rho_1(t_0) - \int_{t_0}^{t} e^{-(t-\tau)\mathcal{L}}\{\rho_0, H_1(\tau)\}d\tau. \tag{6.7}$$

3. At $t = t_0$ the electric field is turned on, so that for $t < t_0, \mathbf{E}(t) = 0$, and $\rho_1(t_0) = 0$.

Then to first order in the electric field, the phase-space density ρ is given as

$$\begin{aligned}
\rho &= \rho_0 + \rho_1 \\
&= \rho_0 - \int_{t_0}^{t} d\tau e^{-(t-\tau)\mathcal{L}} \left\{ \frac{1}{Z} e^{-\beta H_0}, -\sum_i q_i [\mathbf{r}_i \cdot \mathbf{E}(\tau)] \right\} \\
&= \rho_0 - \int_{t_0}^{t} d\tau e^{-(t-\tau)\mathcal{L}} \times \\
&\quad \sum_i \left[\frac{\partial}{\partial \mathbf{r}_i} \frac{1}{Z} e^{-\beta H_0} \frac{\partial H_1}{\partial \mathbf{p}_i} - \frac{\partial H_1}{\partial \mathbf{r}_i} \frac{\partial}{\partial \mathbf{p}_i} \frac{1}{Z} e^{-\beta H_0} \right]. \quad (6.8)
\end{aligned}$$

The first term in the brackets on the right-hand side is zero, because H_1 doesn't depend on $\mathbf{p}_i$, but the second term gives

$$\rho = \rho_0 - \int_{t_0}^{t} e^{-(t-\tau)\mathcal{L}} \left(-\sum_i -q_i \mathbf{E}_i(\tau) \cdot (-\frac{\beta \mathbf{p}_i}{m}) \rho_0 \right) d\tau. \quad (6.9)$$

The microscopic form of the total electrical current $\mathbf{j}$ is defined by $\mathbf{j} = \sum_i q_i \mathbf{v}_i$, so that

$$\rho = \rho_0 + \beta \int_{t_0}^{t} e^{-(t-\tau)\mathcal{L}} \rho_0 \mathbf{j} \cdot \mathbf{E}(\tau) d\tau. \quad (6.10)$$

We take a slight digression in order to understand this result. To do this, we consider some function in phase-space that has no explicit time dependence, $A = A(\mathbf{r}_i, \mathbf{p}_i)$. Then A will still change with time since it is a function of the coordinates and momenta, and these quantities do change with time along a trajectory in phase-space. Therefore, the time dependence of A is determined by

$$\frac{dA}{dt} = \sum_i \frac{\mathbf{p}_i}{m} \cdot \frac{\partial A}{\partial \mathbf{r}_i} + \sum_i \dot{\mathbf{p}}_i \cdot \frac{\partial A}{\partial \mathbf{p}_i} = \mathcal{L} A, \quad (6.11)$$

so

$$A(\Gamma_t) = e^{t\mathcal{L}} A(\Gamma). \quad (6.12)$$

The operator $e^{t\mathcal{L}}$ is a time–displacement operator that moves a phase point forwards in time. This can be used to write (6.9) as

$$\rho = \rho_0 + \beta \int_{t_0}^{t} d\tau \mathbf{j}[-(t-\tau)] \cdot \mathbf{E}(\tau) \rho_0[-(t-\tau)].$$

Here, we have used a short-hand notation where $A[-(t-\tau)] \equiv A(\Gamma_{-(t-\tau)})$. Because $H_0(\Gamma)$ is constant in time, it follows that $\rho_0(\Gamma)$ is independent of time and can be taken outside the integral, so that

$$\rho(\Gamma,t) = \rho_0(\Gamma)\left\{1+\beta\int_{t_0}^{t} d\tau \mathbf{E}(\tau)\cdot\mathbf{j}[-(t-\tau)]\right\}. \quad (6.13)$$

With this result for the phase-space distribution function, we can express the ensemble average of $\mathbf{j}$ at time t to first order in the electric field as

$$\begin{aligned}\langle\mathbf{j}\rangle_t &= \int d\Gamma \mathbf{j}(\Gamma)\rho(\Gamma,t)\\ &= \int d\Gamma \mathbf{j}(\Gamma)\rho_0 \times\\ &\qquad \left\{1+\beta\int_{t_0}^{t} d\tau e^{-(t-\tau)\mathcal{L}}[\mathbf{E}(\tau)\cdot\mathbf{j}(\Gamma)]\right\}. \quad (6.14)\end{aligned}$$

The first term is the average electrical current in the absence of an electric field. Written out in full this term looks like

$$\int d\Gamma(\sum_j q_j\mathbf{v}_j)\frac{1}{Z}e^{-\beta[\sum_i \mathbf{p}_i^2/2m+\Phi]} = 0,$$

because each term in the integrand is odd under $\mathbf{v}_j \to -\mathbf{v}_j$, so that in equilibrium there is no current.

In the second term, we change the integration variable to $\tau' = t-\tau$, so that (6.12) can be used to obtain

$$\begin{aligned}\langle\mathbf{j}\rangle_t &= \beta\int d\Gamma\rho_0(\Gamma)\mathbf{j}(\Gamma)\int_0^{t-t_0} d\tau' e^{-\tau'\mathcal{L}}(\mathbf{E}(t-\tau')\cdot\mathbf{j}(\Gamma))\\ &= \beta\int_0^{t-t_0} d\tau \,\langle\mathbf{j}(\Gamma)\mathbf{j}(\Gamma_{-\tau})\rangle\cdot\mathbf{E}(t-\tau). \quad (6.15)\end{aligned}$$

The angular brackets denote the equilibrium canonical ensemble average. We take $\mathbf{E}(t)$ to be an oscillating field, $\mathbf{E}(t) = \mathbf{E}_0 e^{i\omega t}$ $(t>0)$, so that we have

$$\langle\mathbf{j}\rangle_t = \beta\int_0^{t-t_0} d\tau e^{-i\omega\tau}\,\langle\mathbf{j}(\Gamma)\mathbf{j}(\Gamma_{-\tau})\rangle\cdot\mathbf{E}_0 e^{i\omega t}.$$

Let $t_0 = 0$ and $t\to\infty$. Then we obtain Ohm's law for $\mathbf{J}_e = \langle\mathbf{j}\rangle_t/V$:

$$\mathbf{J}_e = \sigma(\omega)\cdot\mathbf{E}(t), \quad (6.16)$$

where $\sigma(\omega)$ is a tensor of rank two,

$$\sigma(\omega) = \frac{\beta}{V}\int_0^{\infty} d\tau e^{-i\omega\tau}\,\langle\mathbf{j}(\Gamma)\mathbf{j}(\Gamma_{-\tau})\rangle, \quad (6.17)$$

which depends on the whole history of the system. This is the famous Green–Kubo formula for the electrical conductivity of a system of charged particles. The average in the integrand is an equilibrium time–correlation function

$$\langle \mathbf{j}(\Gamma)\mathbf{j}(\Gamma_t)\rangle = \frac{1}{Z}\int d\Gamma e^{-\beta H_0}\sum_i\sum_j q_i q_j \mathbf{v}_i(0)\mathbf{v}_j(t). \qquad (6.18)$$

Equation (6.12), written out explicitly, is

$$A(\Gamma_t) = \exp\left[t\sum_i\left(\mathbf{v}_i\cdot\frac{\partial}{\partial\mathbf{r}_i} + \mathbf{F}_i\cdot\frac{\partial}{\partial\mathbf{p}_i}\right)\right]A(\Gamma), \qquad (6.19)$$

which shows that the combination of $\mathbf{v}\to-\mathbf{v}$ and $t\to-t$ is a symmetry. Since the time–correlation-function is symmetrical under $\mathbf{v}\to-\mathbf{v}$, we also have

$$\langle \mathbf{j}(\Gamma)\mathbf{j}(\Gamma_t)\rangle = \langle \mathbf{j}(\Gamma)\mathbf{j}(\Gamma_{-t})\rangle\,. \qquad (6.20)$$

Finally, the electrical conductivity can be written as

$$\sigma(\omega) = \frac{\beta}{V}\int_0^\infty d\tau e^{-i\omega t}\,\langle \mathbf{j}(\Gamma)\mathbf{j}(\Gamma_t)\rangle\,. \qquad (6.21)$$

This formula is generally believed to be correct, and it is in agreement with experimental and computer simulation results for small electric fields. Moreover, it is possible to derive similar expressions for all of the transport coefficients – viscosities, thermal conductivities, etc. – by devising external fields that induce shearing in a fluid, or temperature gradients. There are also alternative derivations which do not rely on external fields but instead are based upon Chapman–Enskog solutions of the Liouville equation, similar in spirit, at least, to the Chapman–Enskog solution of the Boltzmann equation presented in Chapter 2.

6.2 van Kampen's objections

N. G. van Kampen has argued that linear-response theory derivations of the Green–Kubo formulae as given above for the electrical conductivity are at best questionable.

While not objecting particularly to the Green–Kubo formulae themselves, van Kampen argues that the linear-response derivation of them is artificial and misses the essential points needed for a precise understanding of the validity of the linear laws.

To understand his concerns, let's summarize the steps in the derivation just presented:

1. We solved Liouville's equation by following the phase-space trajectory through every point in Γ-space backward in time.
2. We assumed that the trajectory in an electric field can be determined by an expansion about the unperturbed trajectory in powers of the electric field $\mathbf{E}$. That is, the full equations of motion for the system in the presence of $\mathbf{E}$ were not solved in the derivation of the Green–Kubo formula, instead the dynamics were only kept to first order in $\mathbf{E}$.
   The $\mathbf{E}$-field accelerates the charged particles which leads to 'ohmic' heating which was not taken into account. This heating is neglected because it is of order $\mathbf{E}^2$.
3. We then obtained Ohm's law $\mathbf{J}_\mathrm{e} = \sigma \mathbf{E}$.

This derivation occasioned two objections from van Kampen:

Objection I: The deviation of the trajectory of a particle from its path can only be approximated by linearization if the electric field is extremely small. If the particle moves freely without colliding, its motion is given by $\mathbf{x}(t) = \mathbf{x}_0 + \mathbf{v}_0 t + q\mathbf{E}t^2/2m$. Even neglecting the collisions (which are, in fact, crucial to our understanding of what is going on here) during a time t, the linearized equation is only correct if the third term is very small with respect to the average distance between particles. An estimate can be found in van Kampen's paper: $|\mathbf{E}| < 10^{-18}$ volts/cm.
Objection II: The derivation suggests that the linear dependence of the macroscopic quantity $\langle J \rangle$ on the electric field originates from a linear effect the field has on the trajectory of a particle, but the actual trajectory of a single particle changes very drastically as a result of the field (see Fig. 6.1): *macroscopic linearity is different from microscopic linearity.*

It is here that the discussion of chaos in the previous chapter becomes very relevant! Remember that two nearby trajectories will eventually move into very different regions of phase-space if the system is mixing. Consider, as an example, the Lorentz model of an electron moving in an array of fixed scatterers placed at random in space. Now suppose that we consider a trajectory starting at some phase point for the electron given by $(\mathbf{r}_0, \mathbf{v}_0)$. Suppose we

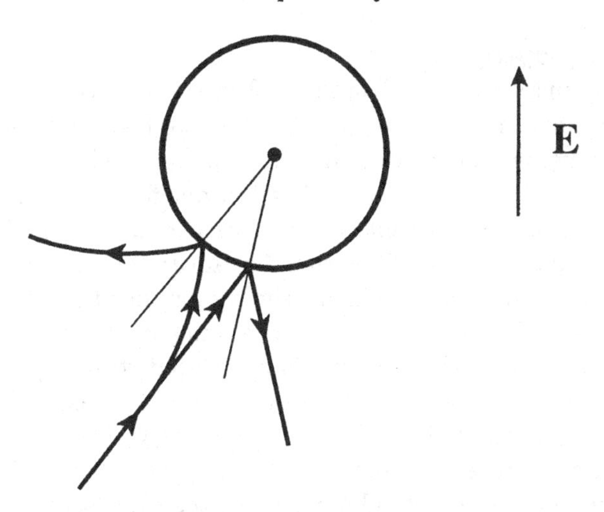

Fig. 6.1 Non-linear divergence of trajectory in a field from a field-free trajectory with identical initial conditions.

compare the trajectory of the electron starting from this point obtained without an external electric field, with the trajectory from this point but when an external electric field is applied to the system. After a few collisions the trajectories will be very different – the field-free and the non-zero field trajectories will be in very different parts of configuration and velocity space. The difference between the two trajectories will not in any sense be able to be described as being linear in the applied electric field, no matter how small the field may be. Thus, van Kampen's objections to this linear response theory derivation of the Green–Kubo formulae is perfectly consistent with our understanding of the chaotic behavior of physical systems in which normal transport processes are taking place. We conclude that a careful derivation of the Green–Kubo formulae requires a linearization in the applied fields to be carried out only *after* the effects of the external fields on the dynamics has been correctly accounted for. The linear laws of hydrodynamics, such as Ohm's law, are not only due to the smallness of the applied field, but must also be due to some very ergodic behavior of the system in the presence of the field, which allows an expansion of the average properties of the system in powers of the field strength when all of the phase-space averaging has been carried out properly.

The responses to van Kampen's objections in the papers and book mentioned in *Further reading* below are based on the following very plausible line of reasoning: Statistical mechanical calculations really refer to probability distributions for the behavior of a large number of independent electrons rather than to the behavior of an individual electron. The crucial quantity for such probability distribution, in any more rigorous theory for the average behavior of the electron, is very likely to be the range of times t for which, $\mathbf{v}(t)$, the velocity of the electron is correlated to its initial velocity, $\mathbf{v}(0)$. We know, both from theoretical studies as well as from computer simulations, that this time interval will be on the order of a mean free time, t_0. One would certainly expect this on the basis of ergodic arguments alone, since after a few scatterings, the particle's velocity becomes randomized. Therefore, the time t that should be inserted in the formula to determine the range of validity in Kubo's argument, $q\mathbf{E}t^2/2m$, is the mean free time t_0, rather than the physical time t. For a dilute gas, t_0 is on the order of 10^{-9} s. This effect alone is sufficient to get the value of $\mathbf{E}$, for which this fraction is small, in the range of physically realistic values. Nevertheless, it is still necessary to refine the mathematical support for the Green–Kubo formulae to the point that these arguments become clearer and more rigorous, rather than just plausible. A really fundamental derivation of these formulae must certainly rely on an ergodic theorem which will allow the linearization of ensemble averages, and will show that Green–Kubo formulae can be applied to systems subjected to realistic external fields.

In the derivation of the equation for the probability current for a tagged particle, $\mathbf{J}_t = -D\nabla P$, from the Boltzmann equation (see Sec. 2.4), we obtained a linear law only after we had used some kind of ensemble averaging to derive the Boltzmann equation, in contrast to the derivation of the Green–Kubo formula, where the ensemble averaging took place after linearization. Green's derivation of the Green–Kubo formulae used a Fokker–Planck equation in phase-space, and had many similarities to Boltzmann's derivation of hydrodynamic equation from the Boltzmann equation. In Green's derivation, the ensemble averages were carried out first and only then were the resulting equations linearized to give the Green–Kubo formulae.

6.3 The Green–Kubo formula: diffusion

Now that we have discussed the linear response derivation of the time–correlation–function expression for the coefficient of electrical conductivity, and the various problems associated with this derivation, we now turn to the standard derivation of the Green–Kubo formulae. This derivation does not rely on perturbation theory for determining the phase-space trajectories of a system. Instead, its roots are in the Chapman–Enskog solution of the Boltzmann transport equation. A nonequilibrium macroscopic state is set up, and we follow the relaxation of this state to an equilibrium state. As a result, any ergodic type theorems that may be required for a rigorous understanding of the method may be formulated in terms of the behavior of certain time–correlation-functions, as we will see presently.

As we did for the Boltzmann equation (see Sec. 2.4), we consider a simple transport process where we study the diffusion of a tagged particle in a fluid of particles that are mechanically identical to the tagged one, and where the untagged 'bath' particles are maintained in a state of thermal equilibrium. More complicated transport processes can and have been discussed in the literature, but this case will suffice to illustrate the method.

We begin by observing that if the fluid were in total equilibrium – including the tagged particle – then the probability of finding the tagged particle in a region of volume $d\mathbf{r}$ about a point $\mathbf{r}$ in the fluid would be $d\mathbf{r}/V$ for a finite system of volume V. We thus consider an initial situation where the tagged particle has a spatial probability distribution $W(\mathbf{r}_1) \neq 1/V$, and we use the subscript 1 to denote the tagged particle. The rest of the fluid is supposed to be in thermal equilibrium at the initial time. The properly normalized N-particle distribution function which satisfies the above conditions at time $t = 0$ is given by

$$\rho(\Gamma, t = 0) = \frac{V}{Z} e^{-\beta H_N} W(\mathbf{r}_1), \tag{6.22}$$

where Z is given by

$$Z = \int d\Gamma e^{-\beta H_N}, \tag{6.23}$$

and

$$\int d\mathbf{r}_1 W(\mathbf{r}_1) = 1. \tag{6.24}$$

After a time t, the phase-space distribution becomes $\rho(\Gamma, t)$, given by

$$\rho(\Gamma, t) = e^{-t\mathcal{L}}\rho(\Gamma, t = 0) = \frac{V}{Z}e^{-t\mathcal{L}}e^{-\beta H_N}W(\mathbf{r}_1). \qquad (6.25)$$

The probability of finding the tagged particle at a point $\mathbf{r}$ at time t, averaged over the configurations and velocities of all of the other particles is then

$$\begin{aligned} P(\mathbf{r}, t) &= \int d\Gamma \delta(\mathbf{r}_1 - \mathbf{r})\rho(\Gamma, t) \qquad (6.26) \\ &= \frac{V}{Z}\int d\Gamma \delta(\mathbf{r}_1 - \mathbf{r})e^{-t\mathcal{L}}e^{-\beta H_N}W(\mathbf{r}_1). \end{aligned}$$

This expression may be transformed using Liouville's theorem to

$$P(\mathbf{r}, t) = \frac{V}{Z}\int d\Gamma e^{-\beta H_N}W(\mathbf{r}_1)\delta(\mathbf{r}_1(t) - \mathbf{r}), \qquad (6.27)$$

where $\delta(\mathbf{r}_1(t) - \mathbf{r}) = \exp(t\mathcal{L})\delta(\mathbf{r}_1 - \mathbf{r})$. It is now useful to insert the Fourier expansion of the delta function in this expression so as to obtain

$$P(\mathbf{r}, t) = \frac{1}{V}\sum_{\mathbf{k}} e^{-i\mathbf{k}\cdot\mathbf{r}}P_{\mathbf{k}}(t), \qquad (6.28)$$

where

$$P_{\mathbf{k}}(t) = \frac{V}{Z}\int d\Gamma e^{-\beta H_N}W(\mathbf{r}_1)e^{i\mathbf{k}\cdot\mathbf{r}_1(t)}. \qquad (6.29)$$

The point of this manipulation is to get the probability distribution in a form where we can follow the time dependence of the long-wavelength component of it. The long-wavelength component is the part that should show hydrodynamic behavior for long times (since it should change on spatial scales on the order of the range of the density gradients), so we need to look at $P_{\mathbf{k}}(t)$ for small $|\mathbf{k}|$. Equation (6.29) can be simplified some more if we remember that both H_N and $[\mathbf{r}_1(t) - \mathbf{r}_1]$ depend only on relative distances $\mathbf{r}_{ij} = \mathbf{r}_i - \mathbf{r}_j$. If we suppose that we place periodic boundary conditions on the system, we can write

$$\begin{aligned} P_{\mathbf{k}}(t) &= \frac{V}{Z}\int d\Gamma e^{i\mathbf{k}\cdot\mathbf{r}_1}W(\mathbf{r}_1)e^{-\beta H_N}e^{i\mathbf{k}\cdot[\mathbf{r}_1(t)-\mathbf{r}_1]} \\ &= W_{\mathbf{k}}F(\mathbf{k}, t), \qquad (6.30) \end{aligned}$$

where

$$W_{\mathbf{k}} = \int d\mathbf{r}_1 e^{i\mathbf{k}\cdot\mathbf{r}_1}W(\mathbf{r}_1),$$

and the *intermediate scattering function* $F(\mathbf{k},t)$ is given by

$$\begin{aligned} F(\mathbf{k},t) &= \frac{1}{Z}\int d\Gamma e^{-\beta H_N} e^{i\mathbf{k}\cdot\Delta\mathbf{r}_1(t)} \\ &= \left\langle e^{i\mathbf{k}\cdot\Delta\mathbf{r}_1(t)}\right\rangle. \end{aligned} \tag{6.31}$$

Here $\Delta\mathbf{r}_1(t) = \mathbf{r}_1(t) - \mathbf{r}_1$, and the angular brackets denote an equilibrium average in the canonical ensemble. The function $F(\mathbf{k},t)$ occurs in the theory of neutron scattering, hence its name.

The function $F(\mathbf{k},t)$ has a particularly convenient form, since it is the generating function for the moments of the displacement of the tagged particle. To obtain a useful expression for $F(\mathbf{k},t)$, we first expand it as a power series in the wave vector $\mathbf{k}$ and then express the result in an exponential, leading to the so-called cumulant expansion. That is

$$F(\mathbf{k},t) = 1 - \frac{k^2}{2}\left\langle\left[\hat{\mathbf{k}}\cdot\Delta\mathbf{r}_1(t)\right]^2\right\rangle + \cdots, \tag{6.32}$$

where $\hat{\mathbf{k}}$ is a unit vector in the direction of $\mathbf{k}$, and $|\mathbf{k}| = k$. All odd powers of $\mathbf{k}$ vanish in this series due to the spatial isotropy of the equilibrium average. We omit the higher terms since we will not be concerned with them here, but it is certainly possible to develop this series further. Equation (6.32) can be written in exponential form as

$$F(\mathbf{k},t) = \exp\left\{-\frac{k^2}{2}\left\langle\left[\hat{\mathbf{k}}\cdot\Delta\mathbf{r}_1(t)\right]^2\right\rangle + \cdots\right\}. \tag{6.33}$$

The quantity appearing in the angular brackets in (6.33) does not depend on the spatial direction of $\hat{\mathbf{k}}$ and we may replace this with the value in any particular direction, the x-direction, say, so that

$$F(\mathbf{k},t) = \exp\left\{-\frac{k^2}{2}\left\langle[\Delta r_{1x}(t)]^2\right\rangle + \cdots\right\}. \tag{6.34}$$

Equations (6.28)–(6.34) can be combined to yield an equation for $P(\mathbf{r},t)$,

$$\begin{aligned} \frac{\partial P(\mathbf{r},t)}{\partial t} &= \frac{1}{V}\sum_{\mathbf{k}} e^{-i\mathbf{k}\cdot\mathbf{r}} W_{\mathbf{k}} \frac{\partial F(\mathbf{k},t)}{\partial t} \\ &= -\left\langle v_{1x}(t)\Delta r_{1x}(t)\right\rangle \frac{1}{V}\sum_{\mathbf{k}} k^2 e^{-i\mathbf{k}\cdot\mathbf{r}} W_{\mathbf{k}} F(\mathbf{k},t) + \cdots \\ &= D(t)\nabla^2 P(\mathbf{r},t) + \cdots, \end{aligned} \tag{6.35}$$

which looks suspiciously like the diffusion equation that we encountered in Chapter 2. The time-dependent diffusion coefficient $D(t)$ in (6.35) is given by

$$\begin{aligned} D(t) &= \langle v_{1x}\Delta r_{1x}\rangle = \int_0^t \langle v_{1x}(t)v_{1x}(\tau)\rangle\, d\tau \\ &= \int_0^t \langle v_{1x}(0)v_{1x}(\tau)\rangle\, d\tau. \end{aligned} \tag{6.36}$$

Here, we have used the fact that the equilibrium average of the product of the velocities of the tagged particle at two different times, t and τ, depends only on the interval $(t-\tau)$. The time-dependent diffusion coefficient is clearly zero at $t=0$. However, we should only expect to obtain a diffusion equation after the tagged particle has had enough time to collide with a number of particles in the fluid such that its trajectory approaches more and more closely that of a particle undergoing a random walk. Thus we expect that $D(t)$ should approach a constant, non-zero value for times long compared to the mean free time between collisions, t_0. If this is so, we can identify the macroscopic diffusion coefficient with the microscopic quantity

$$D = \lim_{t\gg t_0} \lim_{N\to\infty, V\to\infty, N/V=n} \int_0^t \langle v_{1x}(0)v_{1x}(\tau)\rangle\, d\tau, \tag{6.37}$$

where we have also taken the thermodynamic limit so as to remove any dependence of the diffusion coefficient on the system size or on the boundary conditions at the walls of the container. We have therefore obtained an expression for the coefficient of tagged particle diffusion as a time integral of the velocity autocorrelation function for the tagged particle. This is an example of the Green–Kubo formulae for the transport coefficients of a fluid.

We also note an interesting relation between the mean square displacement of the tagged particle and the diffusion coefficient, namely,

$$\begin{aligned} \frac{d}{dt}\left\langle [\Delta r_{1x}(t)]^2\right\rangle &= \frac{d}{dt}\int_0^t dt_1 \int_0^t dt_2\, \langle v_{1x}(t_1)v_{1x}(t_2)\rangle \\ &= 2\int_0^t d\tau\, \langle v_{1x}(0)v_{1x}(\tau)\rangle \\ &= 2D(t) \\ &= 2D \ \text{ for } t \gg t_0. \end{aligned} \tag{6.38}$$

We therefore have arrived at the result announced at the end of Chapter 2, that for large times the mean square displacement of the tagged particle satisfies the relation

$$\left\langle [\Delta r_{1x}(t)]^2 \right\rangle = 2Dt. \tag{6.39}$$

The evaluation of time–correlation-functions for systems of physical interest is an entire subject in itself, and of considerable interest to researchers in nonequilibrium statistical mechanics. The references listed below will provide the interested reader an introduction to this subject.

One of the interesting subjects for research in this area is to establish the precise behavior of the time–correlation-functions using methods of dynamical systems theory, and to compare those results with calculations based on the kinetic theory of gases or with the results of computer simulations.

6.4 Further reading

A survey of the Green–Kubo (or the time–correlation–function) method for transport coefficients with references can be found in the books of Boon and Yip [BY91], Berne [Ber77] and in the article of Dorfman and van Beijeren [DvB77]. The classic paper of van Kampen objecting to Kubo's derivation of the Green–Kubo formulae appears in *Physica Norvegica* [vK71]. Responses to van Kampen's objections can be found in the book of Kubo *et al.* [KTH91], and in papers by Morriss *et al.* [MECvB89], and Chernov *et al.* [CELS93].

Recent rigorous results of Ruelle [Rue97a] and of Gallavotti and Ruelle [GR97] concerning the differentiation of nonequilibrium distribution functions, or more correctly, SRB measures, with respect to a parameter, are very much in the spirit of Kubo's derivation of the Green–Kubo formulae. This recent work makes heavy use of the chaotic properties of smooth Anosov systems, and SRB measures, topics which will be discussed in further chapters, here. Thus, Ruelle and Gallavotti have provided a rigorous derivation of the Green–Kubo formulae and related results, such as the Onsager symmetry relations, for a class of highly chaotic systems with 'nice' mathematical properties. A related discussion of the Green–Kubo formulae based upon the Gallavotti–Cohen fluctuation formula, discussed in Chapter 13, has been provided by Gallavotti [Gal96]. Some earlier, related work on the change of chaotic properties, in this case entropies, with respect to a small perturbation is given in the paper of Katok and co-workers [KKPW90].

7
The baker's transformation

7.1 The transformation and its properties

We return to our discussion of dynamical systems, and consider an example of great illustrative value for the applications of chaos theory to statistical mechanics, the baker's transformation. For this example, we take the phase-space to be a unit square in the (x, y)-plane, with $0 \leq x, y \leq 1$. The measure-preserving transformation will be an expansion in the x-direction by a factor of 2 and a contraction in the y-direction by a factor of $1/2$, arranged in such a way that the unit square is mapped onto itself at each unit of time.

The transformation consists of two steps (see Fig. 7.1): First, the unit square is contracted in the y-direction and stretched in the x-direction by a factor of 2. This doesn't change the volume of any initial region. The unit square becomes a rectangle occupying the region $0 \leq x \leq 2$; $0 \leq y \leq 1/2$. Next, the rectangle is cut in the middle and the right half is put on top of the left half to recover a square. This doesn't change volume either. This transformation is reversible except on the lines where the area was cut in two and glued back.

To write an expression for the baker's transformation, $(x, y) \rightarrow (x', y') = B(x, y)$, we need to distinguish between $x < 1/2$ and $x \geq 1/2$:

$$(x', y') = B(x, y) = \begin{cases} (2x, y/2) & \text{for } x < 1/2 \\ (2x - 1, (y + 1)/2) & \text{for } x \geq 1/2 \end{cases} . \qquad (7.1)$$

The inverse of the baker's transformation, $(x, y) \rightarrow (x', y') = B^{-1}(x, y)$ depends on whether $y < 1/2$ or $y \geq 1/2$:

$$(x', y') = B^{-1}(x, y) = \begin{cases} (x/2, 2y) & \text{for } y < 1/2 \\ ((x + 1)/2, 2y - 1) & \text{for } y \geq 1/2 \end{cases} . \qquad (7.2)$$

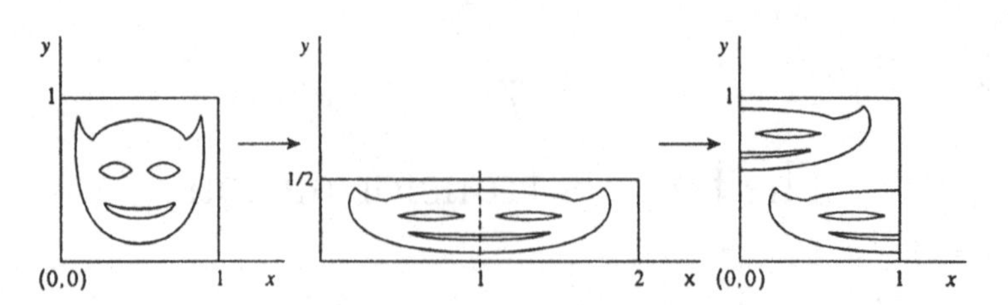

Fig. 7.1. The baker's transformation.

We will show in Chapter 8 that the baker's transformation is ergodic and mixing. The proof of ergodicity is a simple example of a more general method which is used to prove ergodicity in more complicated situations. It is also worth mentioning that, due to the discontinuity in the map, the baker's transformation is not a *diffeomorphism*. That is, it is not an example of a continuous and continuously differentiable map with a continuous and continuously differentiable inverse. While the discontinuity of the baker's map causes no problems with the applications we discuss here, one has to be careful when trying to apply mathematical results which depend in a crucial way on the map being a diffeomorphism.

7.2 A model Boltzmann equation

It is possible to derive a 'Boltzmann equation' for the time-reversible baker's transformation and to show that an H-theorem holds for this equation. Here, then, is one example where the program of Boltzmann can be carried out in detail. The price that we pay for the simplicity of the model is a lack of physical motivation. We will have to supply some of that as we go along. We can begin here by considering the baker's transformation either as a very simple model where we take the two phase-space dimensions to represent, metaphorically, the $2Nd$ dimensions of a large system, or perhaps, more properly, as the projection of a full phase-space onto a subspace of low dimensions where the projected dynamics is ergodic and mixing.

Consider a density function $\rho(x, y)$ on the unit square, that satisfies a Liouville equation for discrete time (the *Frobenius–Perron equation*):

$$\rho_n(x,y) = \rho_{n-1}(B^{-1}(x), B^{-1}(y)),$$

where

$$\rho_n(x,y) = \begin{cases} \rho_{n-1}(x/2, 2y) & \text{for } y < 1/2 \\ \rho_{n-1}((x+1)/2, 2y-1) & \text{for } y > 1/2. \end{cases}$$

Define a reduced distribution function that depends on x only:

$$\begin{aligned} W_n(x) &= \int_0^1 \rho_n(x,y)dy \\ &= \int_0^{\frac{1}{2}} \rho_{n-1}\left(\frac{x}{2}, 2y\right) dy \\ &\quad + \int_{\frac{1}{2}}^1 \rho_{n-1}\left(\frac{x+1}{2}, 2y-1\right) dy. \end{aligned} \tag{7.3}$$

Change to a variable $y' = 2y$ in the first integral and to $y' = 2y-1$ in the second integral. Then,

$$\begin{aligned} W_n(x) &= \frac{1}{2}\int_0^1 dy' \left[\rho_{n-1}\left(\frac{x}{2}, y'\right) + \rho_{n-1}\left(\frac{x+1}{2}, y'\right)\right] \\ &= \frac{1}{2}\left[W_{n-1}\left(\frac{x}{2}\right) + W_{n-1}\left(\frac{x+1}{2}\right)\right]. \end{aligned} \tag{7.4}$$

This is the model Boltzmann equation that is associated with the baker's transformation. We notice that the time is discrete rather than continuous, and that we have selected the x-coordinate for some reason that is not yet clear. It is easy to check that if W_n does not depend on x then W_n remains constant in time. Thus there is an equilibrium distribution W^0 = constant, which corresponds to a uniform distribution on the unit x-interval.

The H-theorem is constructed in the same way as is done for the Boltzmann equation itself. We define $H_n = \int_0^1 dx W_n(x) \ln[W_n(x)]$. Then H develops in time as

$$\begin{aligned} H_{n+1} &= \int_0^1 dx \frac{1}{2}\left[W_n\left(\frac{x}{2}\right) + W_n\left(\frac{x+1}{2}\right)\right] \times \\ &\quad \ln\left\{\frac{1}{2}\left[W_n\left(\frac{x}{2}\right) + W_n\left(\frac{x+1}{2}\right)\right]\right\}. \end{aligned} \tag{7.5}$$

The integrand is of the form $F((a+b)/2)$, where $F(y) = y\ln(y)$. This is a convex function, meaning that a straight line connecting two points of this function is above the function itself (see Fig. 7.2). Therefore $[F(a) + F(b)]/2 \geq F[(a+b)/2]$.

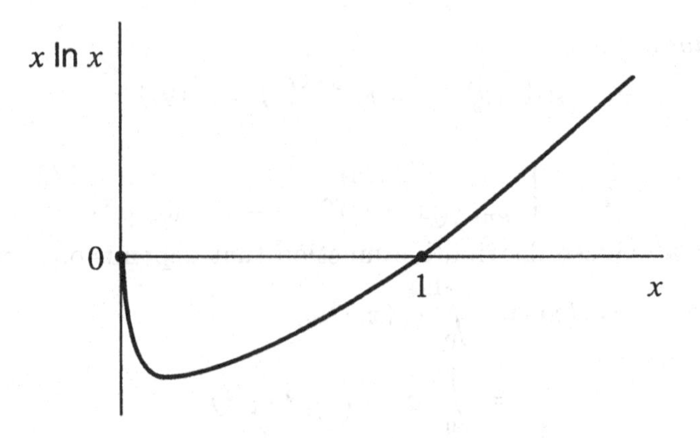

Fig. 7.2. The convex function $x \ln x$.

If we set $a = W_n(x/2)$, and $b = W_n((x+1)/2)$, we have

$$H_{n+1} \leq \frac{1}{2}\int_0^1 dx\left\{W_n\left(\frac{x}{2}\right)\ln\left[W_n\left(\frac{x}{2}\right)\right] + W_n\left(\frac{x+1}{2}\right)\ln\left[W_n\left(\frac{x+1}{2}\right)\right]\right\}.$$

Changing variables to $x' = x/2$ in the first term, and to $x' = (x+1)/2$ in the second term, we find

$$H_{n+1} \leq \int_0^{\frac{1}{2}} dx' W_n(x') \ln[W_n(x')] + \int_{\frac{1}{2}}^1 dx' W_n(x') \ln[W_n(x')]. \tag{7.6}$$

That is, we obtain an H-theorem in the form

$$H_{n+1} \leq H_n. \tag{7.7}$$

We note that H stays constant if W_n is constant and, as mentioned above, this is the equilibrium distribution function. Thus, the baker's transformation allows us to mimic many of the important features of the Boltzmann equation in a very simple time reversible model. Perhaps then we will be able to arrive at some deeper understanding of the Boltzmann equation itself.

First, let's try to solve this model Boltzmann equation. We do this by assuming that W_n is sufficiently regular in x that it can be written as a Fourier series

$$W_n(x) = \sum_j a_j(n) e^{2\pi ijx}. \tag{7.8}$$

Then,

$$\begin{aligned} W_{n+1}(x) &= \frac{1}{2}\left[W_n\left(\frac{x}{2}\right) + W_n\left(\frac{x+1}{2}\right)\right] \\ &= \frac{1}{2}\sum_j a_j(n)(e^{2\pi ijx/2} + e^{2\pi ij(x+1)/2}) \\ &= \frac{1}{2}\sum_j a_j(n) e^{2\pi ijx/2}(1 + e^{\pi ij}). \end{aligned} \tag{7.9}$$

The summand is zero for j odd, so that

$$W_{n+1}(x) = \sum_{j=\text{even}} a_j(n) e^{2\pi ijx/2}. \tag{7.10}$$

At each time step, n, the a_js corresponding to $j = 2^{n-1}k$, where k is odd, disappear from the summation. One can easily convince oneself that in the limit that $n \to \infty$, the only a_j that survives is a_0. This then shows that any initial distribution will approach a uniform, equilibrium distribution as $n \to \infty$. In fact the rate of decay to equilibrium is on the order of 2^{-n}. It is important to note that under typical circumstances, $W_n(x)$ will approach its equilibrium value on time-scales which are typically shorter than those needed for individual trajectories to explore the full phase-space, here, the unit square, or for a small initial region of phase-space to get 'mixed' in the full space. Here we have the first example of a situation where a projected distribution function approaches its equilibrium value on a shorter time-scale than that needed for mixing or ergodicity. We will see further examples in Chapter 17.

Before we turn to a deeper mathematical analysis of the baker's transformation we might mention some features of the x and y variables that we should expect to see in this analysis based on physical arguments. Reverting for the moment to a physical system, a dilute gas, we know that the N-particle distribution function $\rho(\Gamma, t)$ is the fundamental distribution function which really determines the behavior of an ensemble of systems not in equilibrium, and that the function which satisfies the Boltzmann equation is the single-particle distribution function $F(\mathbf{r}, \mathbf{v}, t)$, obtained by integrating over the variables of all but one of the particles.

Thus, we might relate our function $\rho_n(x,y)$ to the phase-space function $\rho(\Gamma,t)$, and the function $W_n(x)$ to the single-particle distribution function $F(\mathbf{r},\mathbf{v},t)$. Recall from our discussion in Chapter 3 that it was Bogoliubov who gave an argument for the special role of this single-particle function in the set of all possible distribution functions that could be obtained by integrating $\rho(\Gamma,t)$. He argued that one can separate rapidly varying functions from slowly varying functions. The functions of physical interest change very slowly in time while the functions of little physical interest change very rapidly in time. The time-scales that Bogoliubov thought were relevant are, in a gas, the duration of a binary collision, the mean free time between collisions, and the time that it takes a particle to travel a macroscopic distance. The irrelevant variables, according to this argument, are those whose characteristic time for variations is the duration of a collision. On this time-scale, the single-particle distribution function stays constant; it varies only on time-scales on the order of the time *between* collisions (as can be seen from the Boltzmann equation). Thus $F(\mathbf{r},\mathbf{v},t)$ should be regarded as a relevant – slowly varying – function, while other possible functions are rapidly varying, and hence irrelevant.

The application of Bogoliubov's arguments to the baker's transformation is that we should expect to find that the x variable is slowly varying in some sense while the y variable is rapidly varying. We will, of course, have to see how this is illustrated in the baker's transformation.

7.3 Bernoulli sequences

The baker's transformation is a reversible, area-preserving, deterministic transformation of the unit square onto itself. It is a very nice caricature of a Hamiltonian mechanical system. Nevertheless, it also possesses all of the random properties of a sequence of tosses of a balanced coin with equal probabilities for 'heads' and 'tails'. It is this somewhat paradoxical randomness property that we expect to see in systems of more physical interest that display an approach to an equilibrium state, i.e., irreversible behavior of hydrodynamic type.

In order to relate the baker's transformation to the tossing of a coin, we note that any point (x,y) in the unit square can be represented by the *dyadic expansion* of the numbers x and y. A dyadic expansion of a number is the expansion in inverse powers of 2 as opposed to the decimal expansion in inverse powers of 10. A number between 0 and 1 can be written with arbitrary precision as

$$x = \frac{a_0}{2} + \frac{a_1}{2^2} + \cdots + \frac{a_n}{2^{n+1}} + \cdots,$$

where $a_i = 0$ or 1. In this way, points on the interval $(0, 1)$ can be represented by an infinite sequence of zeroes and ones, with a '.' at the left end of the sequence, as

$$x = .a_0 a_1 \cdots a_n \cdots. \tag{7.11}$$

For y, we use a slightly different notation,

$$y = .b_{-1} \cdots b_{-n} \cdots, \tag{7.12}$$

with $b_{-i} = 0, 1$, so that

$$y = \frac{b_{-1}}{2} + \frac{b_{-2}}{2^2} + \cdots. \tag{7.13}$$

We can combine the two coordinates into one bi-infinite sequence as

$$(x, y) = \cdots b_{-3} b_{-2} b_{-1} . a_0 a_1 a_2 \cdots, \tag{7.14}$$

with a '.' that separates the x-sequence from the y-sequence.

The baker's transformation now can be represented as a simple shift on this bi-infinite sequence.† To see how this works consider a point in the left side of the unit square, for which $x < 1/2$, then $(x, y) \to (x', y')$, where

$$x' = 2x = 2\left[\frac{a_0}{2} + \frac{a_1}{2^2} + \cdots\right] = \frac{a_1}{2} + \frac{a_2}{2^2} + \cdots, \tag{7.15}$$

because for $x < 1/2$, $a_0 = 0$ and

$$y' = \frac{y}{2} = \frac{b_{-1}}{2^2} + \frac{b_{-2}}{2^3} + \cdots = \frac{0}{2} + \frac{b_{-1}}{2^2} + \cdots. \tag{7.16}$$

Therefore, the sequence

$$(x, y) = \cdots b_{-3} b_{-2} b_{-1} . 0 a_1 a_2 \cdots$$

becomes

$$(x', y') = \cdots b_{-3} b_{-2} b_{-1} 0 . a_1 a_2 \cdots.$$

† That is, the baker's transformation is isomorphic to a shift on the bi-infinite sequence.

Similarly, for $x > 1/2$ we find that the sequence

$$(x, y) = \cdots b_{-3}b_{-2}b_{-1}.1a_1a_2\cdots$$

becomes

$$(x', y') = \cdots b_{-3}b_{-2}b_{-1}1.a_1a_2\cdots.$$

The baker's transformation is thus represented in this way as a shift of the sequence one place to the left, with respect to the '.'.

Every point (x, y) in the unit square can thus be seen as a bi-infinite coin-tossing experiment, where at each time step a zero or one (heads or tails) passes the '.'. The central consequence of this representation is the following: Suppose we have two points on the square that are 'close' to each other. By this we mean that the distance between them is less than some large inverse power of 2, say 2^{-N}. This means that the first N digits on either side of the '.' are the same, but that the two points may differ in the $(N+1)$-st digits. Now, if two points differ only in the $(N+1)$-st digit, initially, this difference will be magnified by a factor of 2 at each time step, so that after N steps the first x-digits are certainly different, and after $2N$ time steps the resulting numbers will have no particular relation to each other. Thus two nearby points eventually wander off to different parts of the unit square. This is exactly what Gibbs had in mind, after all, for phase-space trajectories!

These sequences of zeros and ones are called *Bernoulli sequences.* The shift as a result of the baker's transformation is called the *Bernoulli shift.* The set of points in the unit square is isomorphic to the set of all Bernoulli sequences, and the baker's map is isomorphic to the Bernoulli shift. Here, then, is a reversible, deterministic, area-preserving transformation that has an irreversible transport equation for a reduced distribution function, $W_n(x)$, and an underlying stochastic-like microscopic dynamics. The importance of this observation for an understanding of the foundations of nonequilibrium statistical mechanics must not be underestimated.

We have constructed a *symbolic dynamics* for the baker's map. That is, all of the points in the unit square can be coded in sequences of two symbols, and the dynamics of the baker's transformation can be coded as a shift of one space to the left for each sequence. The representation of a dynamical system as a symbolic dynamics is a very powerful tool which can be used to determine an invariant measure on the symbol sequences, generally known as a measure on *cylinder sets.* The cylinder sets are the sets of sequences with a finite number of identical symbols in the same places in the sequences. For example, the set of all sequences with the general structure $\cdots 0.1 \cdots$ forms a cylinder set. As we see in this example, this particular cylinder set corresponds to a region in the unit square with $x > 1/2, y < 1/2$. Consequently, a measure on a cylinder set corresponds to a measure on the space to which the sequences are isomorphic. Much more can be done with symbolic dynamics, since it allows the representation of a complicated dynamical system is terms of the properties of sequences of a finite number of symbols.

7.4 Further reading

The baker's transformation is one of the most often cited models of a deterministic dynamical system that displays stochastic-like properties. Good discussions can be found in the books of Ott [Ott92], Reichl [Rei98], Arnold and Avez [AA68], and in the article of Berry [Ber87]. A good introduction to symbolic dynamics is contained in the first chapter of the book of Katok and Hasselblatt [KH95].

7.5 Exercises

1. **Slightly generalized baker's transformation.** Consider the following generalization of the baker's transformation. Divide up a unit square into two pieces I and II, of areas α and β respectively, such that $\alpha + \beta = 1$. In I, stretch x by a factor $1/\alpha$ and contract y by α. In II, stretch x by a factor $1/\beta$ and contract y by β. Put II on top of I. This process is repeated many times, so that the original unit square gets mixed throughout the entire square (see Fig. 7.3).

   (a) Write down the equations of motion for this transformation and for its inverse.
   (b) Derive the Liouville equation for this system.
   (c) Integrate over y and derive a Boltzmann equation.

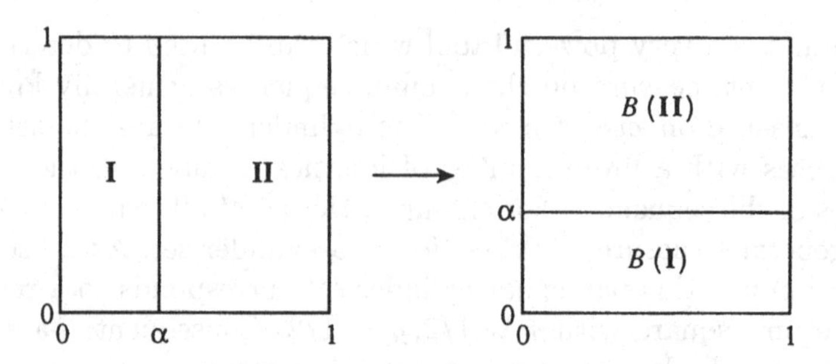

Fig. 7.3. A slightly generalized baker's transformation.

(d) Show that the Boltzmann equation has an H-theorem and an equilibrium solution.

2. Using the dyadic representation of a point (x, y) in the unit square:

(a) Find all points of period 1, 2, 3 and 4 under the baker's transformation. A point of period n satisfies

$$(x, y) = B^n(x, y). \tag{7.17}$$

(b) How many points of period n are there?

3. The version of Liouville's equation for a discrete map M is called the Frobenius–Perron equation, and can be written as

$$\rho_n(\Gamma) = \int d\Gamma' \rho_{n-1}(\Gamma')\delta(\Gamma - M(\Gamma')), \tag{7.18}$$

where n denotes the nth step and Γ a phase point.

(a) Use this to derive the Liouville equation for the baker's transformation.

(b) Consider the one-dimensional *tent map* $M(x)$ (Fig. 7.4, *overpage*):

$$M(x) = \begin{cases} 2x & \text{for } 0 < x < 1/2 \\ 2(1-x) & \text{for } 1/2 < x < 1. \end{cases} \tag{7.19}$$

Find $\rho_{n+1}(x)$ in terms of $\rho_0(x)$. What happens if $n \to \infty$?

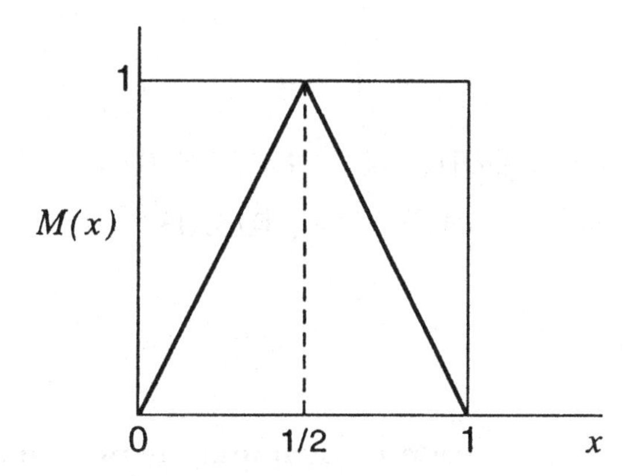

Fig. 7.4. The tent map.

8

Lyapunov exponents, baker's map, and toral automorphisms

8.1 Definition of Lyapunov exponents

In Chapter 7, we have shown that two infinitesimally close points on the unit square can separate from each other under the operation of the baker's transformation. We now want to make this idea of the separation of two trajectories more precise, and to find a mathematical way of computing the rate of separation.

Consider, then, a differentiable (almost everywhere, except, possibly, for a set of measure zero) map M on the interval $(0, 1)$, and examine a small interval $(x_0, x_0 + \delta x_0)$, say. M maps this interval to $(M(x_0), M(x_0+\delta x_0)) = (x_1, x_1+\delta x_1)$, then to $(M(x_1), M(x_1+\delta x_1)) = (x_2, x_2 + \delta x_2)$, etc. After a large number, n, of steps the interval of length δx_0 may become exponentially large or small, that is $\delta x_0 \to \delta x_n = \delta x_0 \exp[n\lambda(x_0)]$, with $\lambda(x_0)$ greater or less than zero. If, after a number of steps, the interval gets folded or cut, then the factor by which the interval is scaled may no longer be approximated in this simple way. Therefore, the limit $\delta x_0 \to 0$ is taken to obtain an expression for $\lambda(x_0)$. Using the above construction, we define a Lyapunov exponent $\lambda(x_0)$ at a point x_0 by

$$\lambda(x_0) = \lim_{n\to\infty} \lim_{\delta x_0 \to 0} \frac{1}{n} \ln \left| \frac{M^n(x_0 + \delta x_0) - M^n(x_0)}{\delta x_0} \right|, \tag{8.1}$$

where $M^n(x) = M(M(M(\cdots M(x)\cdots)))$, that is, the nth iterate of the map at the point x.

The Lyapunov exponent can be written more simply as a derivative,

$$\lambda(x_0) = \lim_{n\to\infty} \frac{1}{n} \ln \left| \frac{dM^n(x_0)}{dx_0} \right|. \tag{8.2}$$

Using the chain rule, we find that

$$\left.\frac{dM^n(x)}{dx}\right|_{x=x_0} = M'(x_{n-1})M'(x_{n-2})\dots M'(x_0),$$

so that a convenient form for the Lyapunov exponent of this map is

$$\lambda(x_0) = \lim_{n\to\infty}\frac{1}{n}\ln\left|\prod_{i=0}^{n-1}M'(x_i)\right| = \lim_{n\to\infty}\frac{1}{n}\sum_{i=0}^{n-1}\ln|M'(x_i)|\,. \quad (8.3)$$

For our purposes, the Lyapunov exponents measure the rate of separation or approach of two nearby phase points as they each follow the trajectory that is determined by the initial value of their phase points. A positive Lyapunov exponent means that two nearby points will separate exponentially with the number of steps, or for a continuous-time map, with time. This property, i.e., the existence of some positive Lyapunov exponents for an isolated dynamical system, is almost certainly needed to understand how the statistical mechanics of irreversible processes follow from the laws of mechanics when they are applied to large systems. The reason for this statement is to be found in the idea of the sensitive dependence of the evolution of a system on its initial state. That is, if a dynamical system has positive Lyapunov exponents, then two nearby initial states will generally separate exponentially in phase space. Thus, we will not be able to predict the outcome of the dynamics unless we know the initial state exactly, and even then, only if we were capable of computing the solutions of the equations of motion to arbitrary accuracy, which we are not. Due to the sensitivity of the dynamical motion upon the initial state, a small range of initial states will lead to a broad range of possible outcomes and a sampling of a large region of phase-space. As a result, successful predictions of the future behavior of the system is likely to be obtained from equations which incorporate some stochastic-like features allowing for a similar sampling of phase-space regions, as in the Boltzmann equation, for example.†

† It is important to note that we usually consider the actions of transformations on compact spaces, such as a unit square, for example. One can see by considering the solution to a simple differential equation like $\dot{x} = kx$, that exponential separation of trajectories may not be very interesting in unbounded spaces.

Thus far, we have presented the definition and construction of Lyapunov exponents for one-dimensional maps. It is also possible to define Lyapunov exponents for higher-dimensional transformations as well, for both maps and flows. This construction can become quite complicated and requires a fairly elaborate and careful formalism for defining the Lyapunov exponents, the expanding directions, and the contracting directions. This structure is provided by what is often referred to as the *multiplicative ergodic theorem* of Oseledets, and is discussed in some detail in the articles by Ruelle and Eckmann, and by Gaspard and Dorfman, (see *Further reading*).

We can briefly outline what is involved in computing the Lyapunov exponents for a flow in a high-dimensional phase-space, but refer the reader to the references for a more detailed discussion. We suppose that a phase point, in configuration and momentum space, is denoted by Γ, and that the Hamilton equations of motion are summarized in the equation

$$\dot{\Gamma}(t) = \mathbf{F}(\Gamma(t)), \tag{8.4}$$

where $\mathbf{F}(\Gamma)$ is a vector which provides the appropriate derivative of the Hamiltonian for each of the coordinates and momenta. We assume that the Hamiltonian does not depend explicitly on time. The solution of the equations of motion then determine the trajectory of the phase point given some initial value $\Gamma(0)$. We are interested in examining the behavior of trajectories with initial values which differ slightly from $\Gamma(0)$; by $\delta\Gamma(0)$, say. The equation of motion for $\delta\Gamma(t)$ can formally be written as

$$\delta\dot{\Gamma}(t) = \mathbf{F}(\Gamma(t) + \delta\Gamma(t)) - \mathbf{F}(\Gamma(t)) \approx \frac{\partial \mathbf{F}}{\partial \Gamma} \cdot \delta\Gamma(t), \tag{8.5}$$

since we are considering only linear deviations about the reference trajectory $\Gamma(t)$. Equation 8.5 has a formal solution in terms of a time-ordered exponential operator $\mathbf{M}(t, \Gamma(0))$, where

$$\begin{aligned} \delta\Gamma(t) &= \mathbf{M}(t, \Gamma(0)) \cdot \delta\Gamma(0), \\ \mathbf{M}(t, \Gamma(0)) &= \mathbf{T} \exp \int_0^t \frac{\partial \mathbf{F}}{\partial \Gamma}, \end{aligned} \tag{8.6}$$

and $\mathbf{T}$ indicates that the exponential operator on the right-hand side of (8.6) is to be time-ordered.

We can now consider the distance, $\|\delta\Gamma(t)\|$, between the reference trajectory, $\Gamma(t)$ and a slight deviation from it, $\Gamma(t) + \delta\Gamma(t)$, given by

$$\|\delta\Gamma(t)\|^2 = \delta\Gamma(0)^{\mathrm{T}} \cdot \mathbf{M}^{\mathrm{T}}(t, \Gamma(0)) \cdot \mathbf{M}(t, \Gamma(0)) \cdot \delta\Gamma(0), \tag{8.7}$$

and the superscript T denotes a transpose. We have now related the distance between the reference trajectory and the deviating trajectory in terms of a real, symmetric matrix $\mathbf{M}^{\mathrm{T}} \cdot \mathbf{M}$ given above in (8.7). This matrix can therefore be diagonalized in terms of its eigenvectors, $\mathbf{u}_i(t, \Gamma(0))$, with corresponding eigenvalues, $\chi_i(t, \Gamma(0))$, as

$$\mathbf{M}^{\mathrm{T}}(t, \Gamma(0)) \cdot \mathbf{M}(t, \Gamma(0)) = \sum_i \mathbf{u}_i(t, \Gamma(0)) \chi_i(t, \Gamma(0)) \mathbf{u}_i^{\mathrm{T}}(t, \Gamma(0)), \tag{8.8}$$

where the number of terms in the sum is equal to the number of dimensions of the phase-space.

The Lyapunov exponents, $\lambda_i(\Gamma(0))$, which in principle may depend on the initial phase point, $\Gamma(0)$, are now defined by

$$\lambda_i(\Gamma(0)) = \lim_{t\to\infty} \frac{1}{2t} \ln \chi_i(t, \Gamma(0)). \tag{8.9}$$

Here the factor $1/2t$ arises because we have essentially defined the Lyapunov exponents in terms of the square of the time evolution matrix $\mathbf{M}$. Positive Lyapunov exponents are, of course, associated with the expanding directions in phase-space, negative ones with contracting directions, and zero Lyapunov exponents with *neutral* directions. Typically the direction in phase-space along the flow is a neutral direction, as we discuss below.

For the baker's transformation, which, as we have seen, does lead to a Boltzmann-like equation, it is easy to see, without a lot of formalities, that there are two Lyapunov exponents. One of them is connected to the expanding direction and has the value $\lambda_+ = \ln 2$. The other Lyapunov exponent is connected to the contracting direction and has the value $\lambda_- = -\ln 2$. For the forward-time operation of the baker's transformation, the expanding direction is along the x-axis, and the contracting direction is along the y-axis. If one considers the time-reversed motion, the expanding and contracting directions change places. Therefore, for the forward motion nearby points separated only in the y-direction approach each other exponentially rapidly, where $\lambda_- = -\ln 2$ is the negative Lyapunov exponent associated with this approach. In the x-direction, points separate exponentially with $\lambda_+ = \ln 2$. For the time-reversed motion, the y-direction becomes expanding and the x-direction becomes contracting.† Note also that the sum of the Lyapunov exponents is zero, $(\lambda_- + \lambda_+ = 0)$ which reflects the fact that the baker's transformation is area-preserving. In other words, if a transformation is area-preserving, the rates of contraction in some directions must be matched by rates of expansion in other directions so that the total area of any region is conserved in the course of the time development of that region.

More generally, one can prove that for symplectic (Hamiltonian) systems all the Lyapunov exponents come in pairs with opposite signs, unless both members of a pair are zero. These $\lambda = 0$ exponents are often connected with conserved quantities. For example, on the constant-energy surface, two points *along* the same individual trajectory do not separate or approach exponentially rapidly, if at all. Thus, there is a zero Lyapunov exponent associated with the direction along a trajectory on the constant-energy surface. The other zero Lyapunov exponent of the pair is then associated with two points in phase-space which lie along a line in a direction perpendicular to the constant-energy surface.

8.2 The baker's transformation is ergodic

We now have all of the tools needed to prove that the baker's transformation B, (7.1), is ergodic. In fact, it is possible to prove a much stronger property of the baker's transformation – it is a Bernoulli process – which implies that it is mixing. A Bernoulli process is one in which it is possible to establish some kind of isomorphism between the process and a random Markov process. This is exactly what we did when we showed that the baker's

† The direction of time in this instance is fixed at the initial state. We could just as easily consider the inverse of the baker's transformation from the beginning, with no change in the logic, only an interchange of x and y as the expanding and contracting directions. If we pursue this logic far enough, we will need to consider the 'big bang' at the origin of the universe as the source of the cosmological and much of the thermodynamic arrow of time. See the discussions by Feynman and by Penrose in the books listed in *Further reading*.

transformation can be mapped onto Bernoulli shifts. Here, we will give the proof that the transformation is ergodic, since this proof is simple and very illustrative of methods often employed in more complicated cases. At the end of this chapter, we will summarize the main points of our discussion of the baker's transformation and its relevance for the theory of irreversible processes.

Consider a small neighborhood of a point (x, y). As discussed above, the vertical line through (x, y) is the contracting direction, also referred to as the *stable manifold* of that point, such that future images of nearby points on this line approach the future images of (x, y) as they travel together with (x, y). On the horizontal line, called the *unstable manifold*, the future images of points move away from future images of (x, y). Under time reversal, the roles of the x- and y-directions are exchanged and the stable manifold becomes unstable, and vice versa. We also know that there is an invariant measure on the unit square, $d\mu = dxdy = dx'dy' = d\mu'$.

Suppose we consider a function $F(\Gamma)$ which is integrable on the unit square Γ, that is, $\int d\mu|F(\Gamma)| < \infty$. Then Birkhoff's theorem can be used. The baker's transformation is ergodic if $\overline{F}(\Gamma)$ is independent of Γ, where $\overline{F}(\Gamma)$ is the infinite-time average of $F(\Gamma_t)$. According to Birkhoff's theorem, the long-time average exists almost everywhere. Now define forward- and backward-time averages,

$$F^+(\Gamma) = \lim_{n\to\infty} \frac{1}{n} \sum_{j=0}^{n-1} F(B^j(\Gamma)), \tag{8.10}$$

$$F^-(\Gamma) = \lim_{n\to\infty} \frac{1}{n} \sum_{j=0}^{n-1} F(B^{-j}(\Gamma)). \tag{8.11}$$

Step 1 of the proof is to show that the forward- and backward-time averages are equal,

$$F^+(\Gamma) = F^-(\Gamma), \tag{8.12}$$

almost everywhere. To do this, we define a set of points A_ϵ where F^+ is larger than F^-, by

$$A_\epsilon = \{\Gamma : F^+(\Gamma) - F^-(\Gamma) > \epsilon\}. \tag{8.13}$$

We have previously shown that the difference between $F^+(\Gamma)$ and $F^+(B\Gamma)$ is zero, since the time average is independent of the particular starting point on a trajectory. Also $F^-(\Gamma) = F^-(B\Gamma)$.

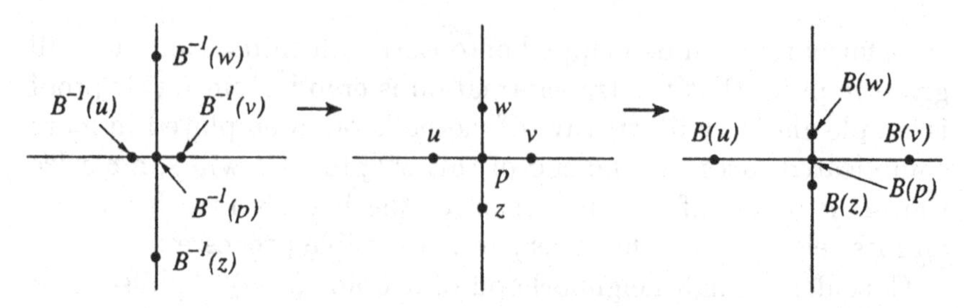

Fig. 8.1 Stable and unstable manifolds for the baker's transformation.

Thus, A_ϵ is an invariant set (i.e., $BA_\epsilon = A_\epsilon$), and then

$$\int_{A_\epsilon} dxdy[F^+(\Gamma) - F^-(\Gamma)] > \epsilon\mu(A_\epsilon). \tag{8.14}$$

Using Birkhoff's theorem, but now applying it to the invariant set A_ϵ, we note that the integral of F^+ over the set A_ϵ is equal to the integral of F over the same region,

$$\int_{A_\epsilon} dxdyF^+(\Gamma) = \int_{A_\epsilon} dxdyF(\Gamma). \tag{8.15}$$

The same holds for the F^- term, so

$$\int_{A_\epsilon} dxdy[F^+(\Gamma) - F^-(\Gamma)] = 0 \geq \epsilon\mu(A_\epsilon). \tag{8.16}$$

This means that $\mu(A_\epsilon) = 0$. Therefore, $F^+(\Gamma) = F^-(\Gamma)$ almost everywhere.

The equality of the forward- and backward-time averages plays a central role in the proof of ergodicity. The idea of the proof is to show that the forward-time average of $F(\Gamma)$ is independent of the y-coordinate of the point $\Gamma = (x, y)$, or $F^+(\Gamma) = F^+(x)$, and that the backward-time average is independent of the x-coordinate of Γ, $F^-(\Gamma) = F^-(y)$. Then, since the forward- and backward-time averages are equal, $F^+(x) = F^-(y)$, and since x and y are independent, we must conclude that $F^+(x) = F^-(y) = \bar{F} =$ constant.

To proceed with the proof, we simplify the notation by replacing the general phase-space notation for a point, Γ, by a simple two-dimensional form $p = (x, y)$. Now, consider a very small square around some point p in the unit square and two points w and z on the stable manifold (see Fig. 8.1).

For two points w and z which are on the same vertical strip (the stable manifold) as the point p, we note that $B^n(w)$ and $B^n(z)$ approach $B^n(p)$ as $n \to \infty$, exponentially. Therefore, the distance between these two points approaches zero exponentially:

$$\begin{aligned} d[B^n(w), B^n(p)] &= \frac{1}{2^n} d(w,p) \to 0, \\ d[B^n(z), B^n(p)] &= \frac{1}{2^n} d(z,p) \to 0. \end{aligned} \tag{8.17}$$

Similarly, two points u and v on the unstable manifold approach p if the system is run backwards, if $n \to \infty$:

$$\begin{aligned} d[B^{-n}(u), B^{-n}(p)] &= \frac{1}{2^n} d(u,p), \\ d[B^{-n}(v), B^{-n}(p)] &= \frac{1}{2^n} d(v,p). \end{aligned} \tag{8.18}$$

Note that we have used the time-reversed motion on the unstable manifold to arrange for points to approach each other and to avoid complications produced by the 'cutting' action of the baker's transformation.

Therefore, if F is a sufficiently smooth function of its variables, then

$$|F(B^n p) - F(B^n w)| \to 0 \tag{8.19}$$

exponentially, as $n \to \infty$, for points w on the stable manifold of p. For points u on the unstable manifold,

$$|F(B^n u) - F(B^n p)| \to 0 \tag{8.20}$$

as $n \to -\infty$. The sum appearing in the time average in (8.10) can be split in such a way that we can use these results:

$$F^+(w) = \lim_{N\to\infty} \frac{1}{N} \left(\sum_{n=0}^{N_\epsilon} + \sum_{N_\epsilon+1}^{N-1} \right) F(B^n w). \tag{8.21}$$

The first sum tends to zero if N_ϵ is fixed at some value. The effect of replacing $F(B^n w)$ in the second sum by $F(B^n p)$ can be made arbitrarily small for large enough N_ϵ. So, for points w on the stable manifold of p, $F^+(w) = F^+(p) = F^+(x)$, where x is the common x-coordinate of the two phase points p and w.

Thus the points on the stable manifold have the same positive-time average, but by the same argument, the points on the unstable manifold have the same negative-time average:

$$F^-(u) = F^-(p) = F^-(v),$$

$$F^+(w) = F^+(p) = F^+(z). \tag{8.22}$$

So, $F^+(p)$ doesn't depend on the y-coordinate of p and $F^-(p)$ doesn't depend on the x-coordinate of p. That is to say, since $\overline{F}(p)$ depends neither on its x-coordinate nor on its y-coordinate, $\overline{F}(p)$ must be a constant, independent of the initial point p. Therefore, the baker's transformation is ergodic.

This method of proving that the forward-time average equals the backward-time average and subsequently showing that the averages on the stable and unstable manifolds depend on different variables and are thus constant, is called Hopf's method, after E. Hopf who first used it to prove that geodesics on surfaces of constant negative curvature are ergodic. In the proof of ergodicity given here we made heavy use of the Lyapunov exponents for the baker's map to ensure the exponential convergence of images of two nearby points, either in the forward or the reversed time directions. It is worth noting that there are ergodic systems with zero Lyapunov exponents, such as the rotations on a circle discussed earlier, and even mixing systems with zero Lyapunov exponents. We refer the reader to the Further readings for examples.

8.3 The baker's transformation and irreversibility

We have now proved that the baker's transformation is ergodic. As a dynamical system, it satisfies the requirements of Boltzmann that the time averages of integrable phase-space functions approach their ensemble averages. Moreover, we can also argue that the system is mixing. Although we will leave the technical details to the reader, the argument is as follows: Consider a small square of length 2^{-N} on a side, inside the unit square. The transformation will stretch this square horizontally and contract it vertically so that after N steps it will be roughly of horizontal dimension unity, and of vertical dimension 2^{-2N}. As the number of steps continues to increase, the original square is transformed into a large number of very thin horizontal strips of length unity, distributed more and more uniformly in the vertical direction. Eventually any small set in the unit square will have the same *fraction* of its area occupied by these little strips as any other set on the square. This is the indicator of a mixing system. A careful proof can be given by using the representation of the baker's transformation as a Bernoulli shift, discussed earlier.

Therefore, the baker's transformation satisfies the postulate of Gibbs and we conclude that a sufficiently smooth initial distribution function defined on the unit square will approach a uniform distribution on the square. We also know that the Liouville equation for the baker's transformation can be integrated to yield a Boltzmann equation for the distribution function on the x-coordinate. If this distribution is sufficiently smooth then we also proved that it approaches a uniform distribution exponentially rapidly in the course of time. There are two points which we now wish to emphasize:

1. The assumption of smoothness is crucial. In Exercise 7.2., we show that there are periodic trajectories on the unit square, of all periods. If the initial distribution on the unit square were to be concentrated on one or more of these periodic orbits, by supposing the initial distribution has the form

$$\rho(\Gamma, n = 0) = \delta(\Gamma - \Gamma_p), \tag{8.23}$$

where Γ_p is some phase point of period p, then $\rho(\Gamma, n)$ would itself be a periodic function with period p and would never approach a uniform distribution in any sense. A similar remark holds for the Boltzmann distribution function $W_n(x)$. There are an infinite number of periodic points for the x-part of the baker's transformation, too. If $W_0(x)$ were concentrated on one or more of these periodic points, $W_n(x)$ would remain concentrated on periodic points and never approach a uniform distribution.

2. When we went from the phase-space distribution function, $\rho(\Gamma, n)$, to a Boltzmann distribution function, $W_n(x)$, it was not obvious why we chose to define our reduced distribution to be a function of the x-coordinate rather than a function of the y-coordinate. We can now understand why this is so. Any small set in phase-space will get smoothed out in the direction of the unstable manifold and get shredded into finer and finer pieces in the direction of the stable manifold. The smoothing will produce functions with a slow variation along the unstable directions, and the shredding will produce functions with very rapid variations along the stable directions. Therefore the stable directions define the rapidly changing variables *in the sense of Bogoliubov*, while the

unstable directions define the slow variables in the same sense. Our job in deriving the Boltzmann equation from the Liouville equation is to use the dynamics to identify the unstable manifolds and then define distributions on the unstable manifolds. These functions will become smoother and smoother as the stretching process takes place, and will eventually approach uniform distributions on the reduced phase-space of the unstable manifolds.

Of course, the baker's transformation takes place on a phase-space of only two dimensions. When we consider an ensemble of systems, we are considering a very large number of copies of the baker's transformation, but each copy is still two-dimensional. This has a consequence that any statistical mechanical analysis of the baker's map will have only two instead of a large number of degrees of freedom. However, there are systems of interest to workers in transport theory that have only a few degrees of freedom, yet they are very important for our understanding of transport. For example, the Lorentz model of an electron moving among fixed scatterers is a common model of diffusion of particles in amorphous (or periodic) media. The baker's map then gives us some insights into the dynamics of such systems. In addition, it might serve as a model that is useful for understanding the projection of a large system onto a space of a few dimensions, as well as serving as a simple system on which to learn some new ideas. We will expand upon these points in Chapter 17, where we continue the discussion of the dynamical foundations of the Boltzmann equation and other related equations that describe irreversible processes in fluids.

Before we continue on to the further issues in dynamical systems theory which we will need for our study of irreversible behavior in deterministic dynamical systems, it is time to introduce a new model system that will be helpful in this study, the Arnold cat map.

8.4 The Arnold cat map

The baker's transformation is a particularly simple example of measure-preserving transformations on the unit square that have the interesting dynamical and stochastic-like properties discussed above. Here, using Figs. 8.2 to 8.6 for illustration, we introduce a slightly more complicated class of models which have similar properties, the so-called *toral automorphisms*, of which the Arnold cat map, Fig. 8.2, to be discussed below is an example.

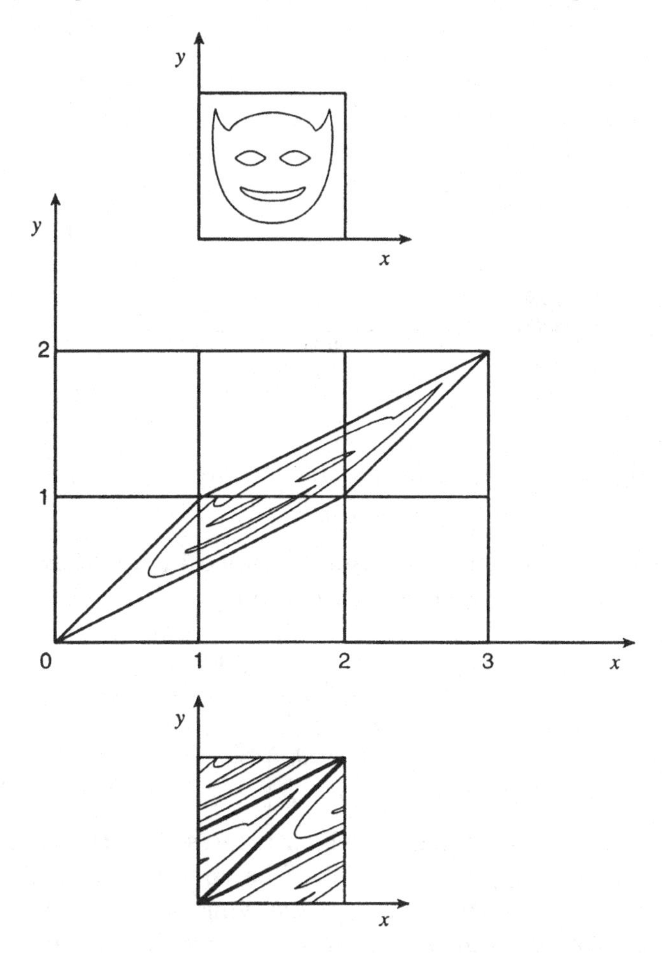

Fig. 8.2. The Arnold cat map.

Toral automorphisms can be defined in two and higher dimensions, but here we consider only the two-dimensional case for simplicity. These are linear maps of the unit square onto itself, with periodic boundary conditions, so that all distances are measured *modulo* 1 in the x- and y-directions, hence, a unit torus is mapped

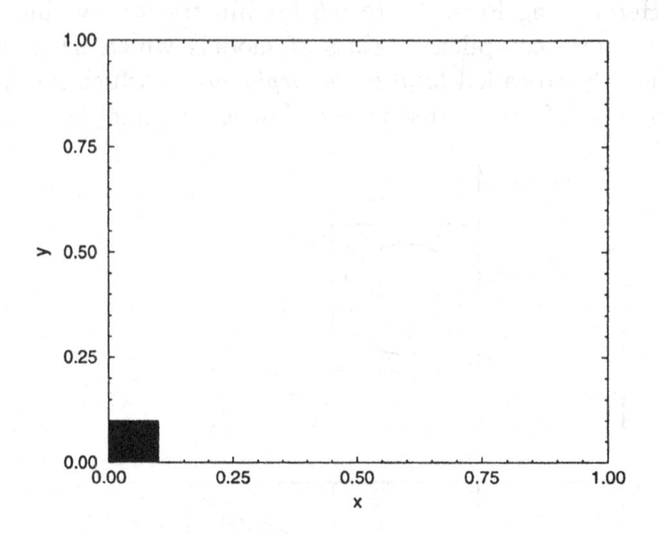

Fig. 8.3. Initial configuration for the Arnold cat map.

onto itself. A toral automorphism is defined by a matrix T, with determinant 1, that describes the mapping

$$\begin{pmatrix} x' \\ y' \end{pmatrix} = T \cdot \begin{pmatrix} x \\ y \end{pmatrix} \pmod 1, \tag{8.24}$$

where

$$T = \begin{bmatrix} a & b \\ c & d \end{bmatrix}. \tag{8.25}$$

Here, a, b, c, d are all integers, which insures that the unit torus will always be mapped onto itself. The condition that $\det T = ad - bc = 1$ ensures the measure-preserving nature of this transformation. We will suppose that T is symmetric – this is not necessary – and that its eigenvalues are real and differ from unity–this is necessary. These conditions guarantee that the eigendirections of T are orthogonal, that there is one eigenvalue larger than 1, and one less than 1. The eigendirection associated with the larger eigenvalue is then the unstable, or expanding direction of the map, and the perpendicular direction associated with the smaller eigenvalue is the stable, or contracting direction. The Lyapunov expo-

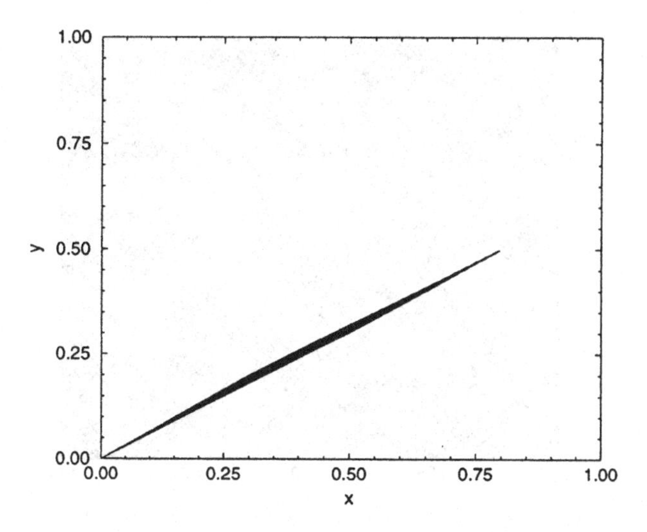

Fig. 8.4 Evolution of the initial configuration after two iterations of the Arnold cat map.

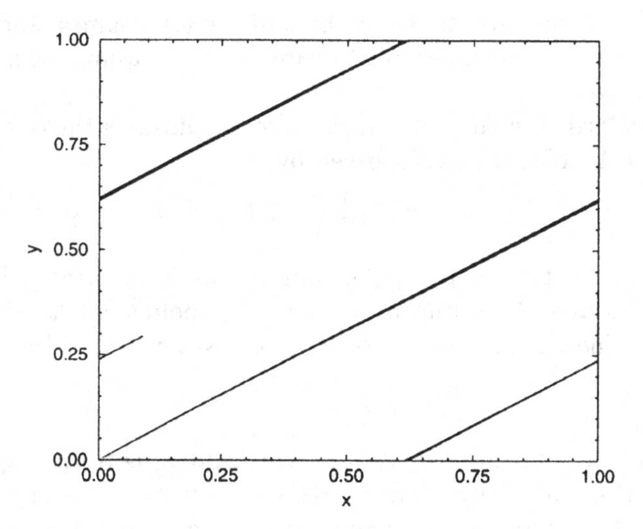

Fig. 8.5 Evolution of the initial configuration after three iterations of the Arnold cat map.

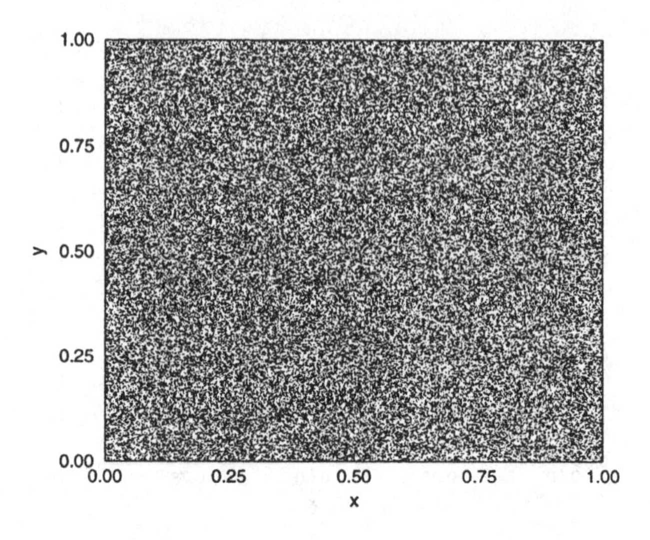

Fig. 8.6 Evolution of the initial configuration after ten iterations of the Arnold cat map.

nents of this map are the logarithms of the eigenvalues, and since the product of eigenvalues must be unity, the Lyapunov exponents form a $\pm$ pair.

A standard version of this toral automorphism is the one introduced by Arnold, where $T_{\rm A}$ given by

$$T_{\rm A} = \begin{bmatrix} 2 & 1 \\ 1 & 1 \end{bmatrix}. \tag{8.26}$$

The action of this map on the unit square is illustrated by the distortions and shredding produced on Arnold's cat as given in Fig. 8.2. The eigenvalues $\Lambda_\pm$ of $T_{\rm A}$ are easily found to be

$$\Lambda_\pm = \frac{3 \pm \sqrt{5}}{2}. \tag{8.27}$$

The stable and unstable directions, corresponding to the eigenvalues less than or greater than 1, respectively, can be easily found as well. The Lyapunov exponents for this map are given by

$$\lambda_\pm = \ln \Lambda_\pm = \ln \left[\frac{3 \pm \sqrt{5}}{2} \right]. \tag{8.28}$$

Now notice what happens if we follow a set of points which are initially confined to a small region on the torus for a few steps. In Fig. 8.3 we illustrate the initial situation, the points are confined to a small square in the lower left-hand corner. Then we show in Figs. 8.4, 8.5, and 8.6 their distribution after two, three, and ten iterations of the map. The original set gets stretched along the unstable manifold and compressed in the stable direction. After ten iterations the points look uniformly distributed in the unit square, but if we were to look carefully enough they would be distributed on about 10^6 lines parallel to the expanding direction and very close together, of course. For this map, we see the same features as we saw in the baker's map, the stretching and folding of an initial set into a smooth set in the unstable direction and an ever increasingly irregular one on finer and finer scales in the stable direction.

Using arguments almost identical to those used for the baker's map, one can prove that the Arnold cat map is ergodic and mixing. That is, for the proof of ergodicity one looks at points aligned along stable or unstable directions. These points approach exponentially in the forward- or backward-time directions, respectively, so that one can easily prove that the time average of an integrable function is a constant on the unit torus. The proof that the cat map is mixing is similar in spirit to that for the baker's map, only the mapping onto a Bernoulli system takes a bit more work and uses the method of *Markov partitions* which we discuss in the next chapter.

Therefore, a smooth enough initial distribution on the unit square will weakly approach an equilibrium distribution on the square. We also have every reason to believe that a suitably projected distribution function will satisfy a Boltzmann-like equation and that an H-theorem will be obtained, just as in the case of the baker's transformation. However, due to the slightly greater complexity of the cat map, we will not be able easily to study these properties analytically. Instead, this map will need to be studied on the computer. However, the reader can easily observe that the projection of the distribution function onto the x- or y-axes will approach a constant value after just a few iterations of the map, while it may take several more iterations of the map before the initial set gets distributed uniformly over the whole unit square. Clearly for this model, as in the baker's map, the positive Lyapunov exponent determines the rate of the approach of the projected distribution function to its equilibrium value.

We will return to the issue of the irreversible behavior of the baker's and cat maps in Chapter 17, but next in Chapter 9 we take up the study of an important dynamical quantity, the *Kolmogorov–Sinai entropy*, associated with the stretching and folding of sets under the action of maps such as the baker's and cat maps.

8.5 Further reading

Lyapunov exponents are central to any discussion of chaotic behavior. Good discussions can be found in the books of Ott [Ott92], Alligood *et al.* [ASY97], and Arnold [Arn89]. A fundamental paper on this subject is due to Wojtkowski [Woj85], and the classic papers of Oseledets and of Pesin can be found in the collection of papers edited by Sinai [Sin91]. See also the discussion in the review article of Ruelle and Eckmann [RE85]. Among the many applications of these ideas to physical systems, the papers of Dorfman *et al.* [DEJ95], and Gaspard and Dorfman [GD95] are useful for our later discussions. The nature of time's arrow and related problems are discussed in the books of Feynman [Fey67], of Penrose [Pen89], and of Schulman [Sch97].

The proof of the ergodic and mixing properties of the baker's map and the Arnold cat map is given in a number of ergodic theory books such as Arnold and Avez [AA68], Katok and Hasselblatt [KH95], Peterson [Pet91], and Robinson [Rob99]. The proof given here is similar to the discussion by Liverani and Wojtkowski [LW95]. The Arnold cat map, toral automorphisms and their various properties are discussed in Arnold and Avez [AA68] and the other books listed above.

8.6 Exercises

1. **One-dimensional random walk.** Construct a *deterministic* one-dimensional map that would provide a dynamical realization of a random walk on a one-dimensional lattice with probability α of going to the right and probability $\beta = 1-\alpha$ of going to the left.

2. Suppose that the map has ergodic properties and compute the Lyapunov exponent for the map.

9
Kolmogorov–Sinai entropy

Now we want to discuss a number of topics that are essential for an understanding of dynamical systems theory and which also play a role in a more detailed discussion of the relation between transport theory and dynamical systems theory. We begin with a discussion of the Kolmogorov–Sinai (KS) entropy, which is a characteristic property of those deterministic dynamical systems with 'randomness' properties similar to Bernoulli shifts discussed earlier. The KS entropy is essential for formulating the *escape-rate* expressions for transport coefficients, to be discussed in Chapters 11 and 12.

9.1 Heuristic considerations

Let us return for a moment to the Arnold cat map discussed in the previous chapter. There we illustrated an initial set, A, say, that is located in the lower left-hand corner of the unit square (see Fig. 8.3). As this set evolves under the action of the map T_{A}, the set becomes longer and thinner so that after three iterations, the set has begun to fold back across the unit square, and after ten iterations, the set is so stretched out that it appears to cover the unit square uniformly (see Figs 8.5 and 8.6, respectively). Since the initial set A is getting stretched along the unstable direction, at every iteration we learn more about the initial location of the points within the initial set A. That is, suppose that we can distinguish two points on the unit square only if they are separated by a distance δ, the resolution parameter, and suppose further that the characteristic dimension of the initial set, A, is of the order of δ. We cannot resolve two points in A then; but after a time, t, the initial set will be stretched along the unstable direction to a length of order $\delta \exp(\lambda_{+}t)$, and we can easily resolve the images of points in the initial set. Thus, if we look at successive images of the initial set we are able to learn more and more about the location of points in the initial region. In fact, this information is growing at an exponential rate. The exponential rate at which information is obtained is measured by the KS entropy, h_{KS}. For the map we are discussing, the Arnold cat map on the unit square (or torus), $h_{\mathrm{KS}} = \lambda_+$. This is an example of Pesin's theorem, which we discuss later in this chapter. Now we turn to a more formal definition.

9.2 The definition of the KS entropy

As emphasized above, and as discussed in some detail by Eckmann and Ruelle, dynamical systems with positive Lyapunov exponents produce information. We now want to apply information theoretic ideas to characterize the production of this information. Consider, then, a phase space or constant-energy surface, Γ, of finite total measure. We suppose that we can decompose Γ into a non-trivial collection of non-overlapping sets $\{W_i\}$ such that

$$\{W_i : \Gamma = \cup_i W_i, W_i \cap W_j = \emptyset \text{ for } i \neq j, \mu(W_i) > 0\}. \tag{9.1}$$

Such a decomposition is called a *partition* of Γ. If we have some transformation B of the phase space onto itself we can characterize many properties of B by considering the future and pre-images of the partition under the action of the transformation B. For concreteness, we will take B to be a discrete-time map which is applied at each time step, as in the baker's transformation.

Given a partition, we can create finer and finer partitions by examining the pre-images of a partition and taking intersections of the original partition with its pre-images. As we will see presently, the sets defined in this way allow us to follow the trajectory of an initial phase point with greater and greater precision as the partitions get finer and finer.

As an example of a partition, we consider a partition of the unit square of the baker's transformation. We take the original partition to be the two sets W_i, with $i = 0, 1$ (see Fig. 9.1). The inverse baker's transformation maps these two sets to $B^{-1}(W_i)$. The intersections of W_i with $B^{-1}(W_j)$ leads to a new partition of

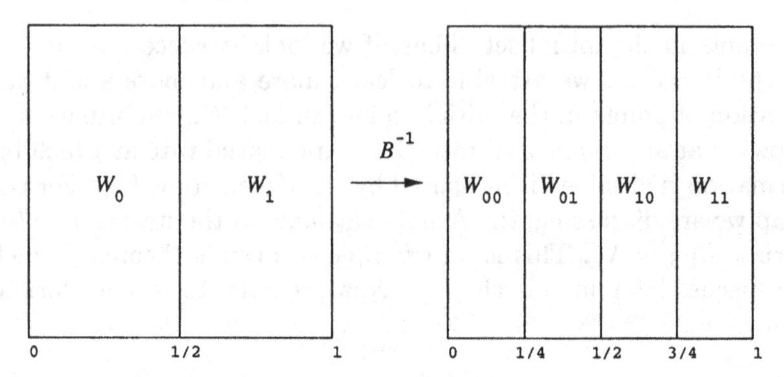

Fig. 9.1 The construction of partitions for the baker's transformation.

the unit square into four sets $W_i \cap B^{-1}(W_j)$, with $i, j = 0, 1$:

$$\begin{aligned} W_{00} &= \{x : x \in W_0, B(x) \in W_0\}, \\ W_{01} &= \{x : x \in W_0, B(x) \in W_1\}, \\ W_{10} &= \{x : x \in W_1, B(x) \in W_0\}, \\ W_{11} &= \{x : x \in W_1, B(x) \in W_1\}. \end{aligned}$$

This partition is the sum (in the mathematical notation, the $\vee$-sum), of the intersections of the partitions $W \equiv \{W_i\}$ and $B^{-1}(W)$. By running the baker's transformation backwards and taking further intersections, finer partitions are obtained, and we get a collection of partitions that contain more and more details about the trajectory of a point:

$$\{\{W_i\},\ \{W_i \cap B^{-1}(W_j)\},\ \{W_i \cap B^{-1}(W_j) \cap B^{-2}(W_k)\},\ \ldots\},$$

or, in the mathematical notation,

$$\{W,\ W \vee B^{-1}(W),\ W \vee B^{-1}(W) \vee B^{-2}(W),\ \ldots\}. \qquad (9.2)$$

To obtain some idea of the meaning of a partition, consider a point x_0 which is in the set $W_0 \cap B^{-1}(W_1) \cap B^{-2}(W_0)$.† Then we know that $x_0 \in W_0$, that $B(x_0) \in W_1$ and that $B^2(x_0) \in W_0$. Thus, by identifying the element of the partition to which a point belongs, one can map out the entire history of a point, as more and more pre-images of the original partition are taken into account. We would like to know how fine the partition is becoming as more

† Note that this sequence of intersections corresponds in a direct way to the symbolic dynamics of the baker's map.

pre-images are included, both as an indication of the mixing of the phase space under the dynamics and of our ability to use larger and larger parts of a trajectory to uniquely specify a particular initial point.

Kolmogorov and Sinai define the entropy of a partition in terms of a normalized invariant measure on the phase space as

$$H(\{W_i\}) = -\sum_i \mu(W_i)\ln(\mu(W_i)), \tag{9.3}$$

with normalization $\sum_i \mu(W_i) = 1$. When the partition is the trivial partition, $W = \Gamma$, then $H = 0$. Then, for the example of the baker's transformation, the partitions in (9.2) have entropies

$$\begin{aligned} H_1 &\equiv H(W) = -\frac{1}{2}(\ln\frac{1}{2} + \ln\frac{1}{2}) = \ln 2, \\ H_2 &\equiv H(W \vee B^{-1}(W)) = -4(\frac{1}{4}\ln\frac{1}{4}) = \ln 4, \\ H_3 &\equiv H(W \vee B^{-1}(W) \vee B^{-2}(W)) = \ln 8, \\ H_{n+1} &\equiv H(W \vee B^{-1}(W) \cdots \vee B^{-n}(W)) = \ln 2^{n+1} \end{aligned}$$

To get a number that indicates how much information is gained per step, define

$$h = \lim_{n\to\infty} \frac{1}{n} H_n. \tag{9.4}$$

This quantity can be interpreted as a measure for the rate at which information is produced for someone who observes the system with a limited resolution. Numerically, it is found that a more useful, but equivalent, definition of h is given by

$$h = \lim_{n\to\infty} [H_{n+1} - H_n]. \tag{9.5}$$

These definitions still depend on the choice of the partition, but the Kolmogorov–Sinai entropy, h_{KS}, is defined as the supremum of the above expression over all possible finite partitions of the space at $t = 0$:

$$h_{\mathrm{KS}} = \sup_{W_i} h. \tag{9.6}$$

A partition that gives the KS entropy directly from (9.4) is called a *generating partition.*

For reasons we shall discuss in the next section, in the case of the baker's transformation, the initial partition chosen above is a generating partition, and leads directly to the maximum value of h, and

$$h_{\mathrm{KS}} = \ln 2.$$

For the rotation map, $x_n \to x_n + \alpha$, on the circle with circumference 1, a partition into two parts will stay a partition into two parts when the system is run backwards. The method described above leads to a partition of the circle into $2n$ intervals after n iterations. So $H(\{W_i\}) = -\sum_i \mu(W_i) \ln \mu(W_i)$ depends logarithmically on n, and the Kolmogorov–Sinai entropy for this map is easily seen to be $h_{\mathrm{KS}} = 0$.

To get a non-zero h_{KS}, the measure at the nth step, $\mu_i = \mu(W_i)$, has to depend exponentially on the number of steps n, so that the average logarithm of that measure, $H(\{W_i\})$, increases linearly with n. The Arnold cat map and the baker's map are examples of what are called *Bernoulli systems.* These systems can be mapped onto Bernoulli sequences with a shift. They are ergodic, mixing, and have positive KS entropy. It is also worth noting that it is possible to define the Kolmogorov–Sinai entropy of a dynamical system without using the notion of partitions, but instead using the notion of an (ϵ, N)-separated set, which will be introduced in Chapter 13 (see *Further reading*).

We note that both the positive Lyapunov exponent and the Kolmogorov–Sinai entropy of the baker's transformation are $\ln 2$. This is an example of Pesin's theorem, which we discuss in the next section.

9.3 Anosov and hyperbolic systems, Markov partitions, and Pesin's theorem

The baker's map and the Arnold cat map on the unit square are simple examples of what are usually referred to as *Anosov hyperbolic systems.* Such a system is defined by the properties that we have found to be important in our analysis of these simple maps. We say that a dynamical system is Anosov, if:

1. The dynamical system is defined by a transformation, T, which we suppose is well-defined for either a discrete-time dynamics or a continuous-time dynamics; an invariant set, Γ, under $T$and an invariant measure, $\mu(A)$, with respect to T, defined on the sets, A,

of Γ. The definition of an invariant measure is very similar to the definition of the invariance of the Liouville measure, and is simply that

$$\mu(A) = \mu(T^{-1}(A)), \tag{9.7}$$

where $T^{-1}(A)$ is the pre-image of A, i.e., the set of points that are mapped into A by the action of T. The pre-image of A is usually well-defined even if T has no unique inverse for points in Γ.
2. The dynamical system is *transitive*. That is, there is at least one trajectory that is dense in the phase-space.
3. At every point, $\mathbf{X}$, in Γ, it is possible to construct stable and unstable manifolds, so that the tangent space at that point can be decomposed into the sum of an *unstable* subspace, a *stable* subspace, and for continuous-time systems, a *center* subspace (which is associated with the direction of the phase flow at that point).
4. The Lyapunov exponents associated with the unstable and stable subspaces are strictly bounded away from zero for all points of Γ with the possible exception of a set of μ-measure zero. The Lyapunov exponent associated with the center manifold for a continuous-time system can be, and usually is, zero.
5. These manifolds vary continuously with the point $\mathbf{X}$, so that stable manifolds do not suddenly turn into unstable ones.
6. These manifolds intersect transversely, with angles that are strictly bounded away from zero. This avoids tangencies which cause a lot of problems and complicate the analysis.

Technically, there is a distinction between an *Anosov system*, which we just defined, and a *hyperbolic system*, which may have these properties only in a subregion of the full space.† However, here we will occasionally ignore this distinction and sometimes refer to systems with the properties 1–5 as hyperbolic systems. It is now clear that the baker's map and the Arnold cat map are examples of Anosov hyperbolic systems. However, the cat map is an example of an *Anosov diffeomorphism*, while, as mentioned earlier in Chapter 7, the baker's map is not a diffeomorphism.‡

† Sometimes the phase space of a system can be decomposed into separate spaces, such that the system is hyperbolic on each of the subspaces. An example occurs in the discussion of Lorentz Lattice Gases in Chapter 16.

‡ Strictly speaking, the baker's map is not an Anosov system because it has a discontinuity at $x = 1/2$, and one usually reserves the term *Anosov system*

A third category of dynamical systems, and one that includes Anosov systems, is called *Axiom-A systems.* Axiom-A systems have a *non-wandering* set of points, Ω,† in the phase-space, such that that on Ω the map is hyperbolic and the fixed and periodic points of the map are dense in Ω. One often encounters systems with *Axiom-A attractors.* Such an attractor will be encountered in our discussion of SRB measures in a subsequent chapter.

The point of introducing these definitions is to provide the reader some starting points for entering the mathematical literature. The reader will note that the Arnold cat map is an Axiom-A system, and that Axiom-A systems are Anosov if Ω coincides with the full phase-space. Here, we will be concerned with the fact that for an Anosov system we can use the method of Markov partitions to compute the KS entropy.

The reader will have noticed that the partitions used in the analysis above for the baker's map had boundaries that coincided with stable and unstable directions of the map. They were examples of what are called Markov partitions. A Markov partition of our space, Γ, is a partitioning of Γ into sets with disjoint interiors, called *rectangles*, whose sides are small, effectively linear or flat, pieces of stable and unstable manifolds. Moreover, the rectangles must be such that when they are transformed by the operations of T and T^{-1}, their images fit nicely within the rectangles of the original partition. That is, under T, the images of the boundaries lying along unstable directions of a rectangle must be stretched so that the transformed rectangle is enclosed within rectangles of the partition, with the images of the boundaries along the stable directions lying along, but enclosed by, the stable boundaries of the original partition. Similarly, under T^{-1}, the sides of the rectangles along the unstable directions are contained in the unstable sides of the original partition, while the images of the stable boundaries

for maps or flows that have no discontinuities. However, for many purposes this is only a technical distinction, and the baker's map can be treated as an Anosov system, or as an Axiom-A system, defined below.

† A wandering set of points in phase-space under the map T is the set W of points with a neighborhood $\mathcal{M}$, such that for large enough n, $\mathcal{M} \cap T^{n'}\mathcal{M} = \emptyset$ for all $n' > n$. All points that are not wandering form the non-wandering set. The reader will note that the Poincaré recurrence theorem states that almost every point in a compact phase space is non-wandering under Hamiltonian dynamics.

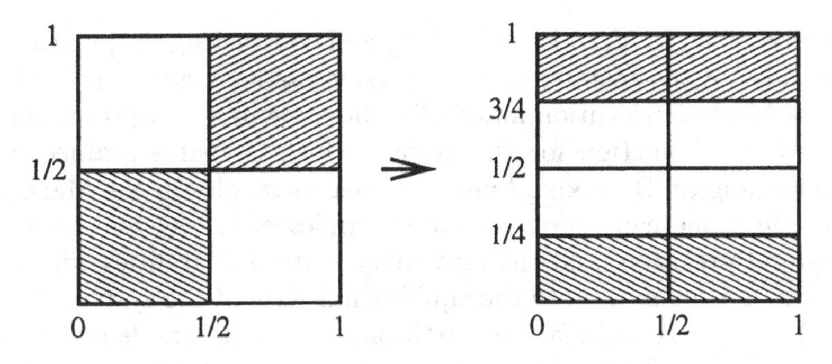

Fig. 9.2. Example of a Markov partition.

are stretched so that the new rectangle lies within rectangles of the original partition. In most cases, though, the boundary sides of the rectangles are not smooth, so the illustrations of Markov partitions for the baker's map and similar figures for two-dimensional toral automorphisms are not at all typical. We refer the reader to *Further reading* for appropriate references.

One can easily check that a partition of the unit square obtained by drawing horizontal and/or vertical lines midway through the square forms a simple Markov partition for the baker's map as illustrated in Fig 9.2. Here we illustrate the transformation of the rectangles under the map B. We can construct the image of the original partition under the inverse map as well. By taking intersections of all of these rectangles created in the forward and inverse maps we can create a very fine partition of the unit square into rectangles, and use these rectangles to represent the dynamics by a Markov process, as discussed below. In general, it is not so easy to construct Markov partitions for dynamical systems, though it can be done without too much trouble for the Arnold cat map.

The main points that we want to emphasize here are that one can prove that it is possible to construct a Markov partition for an Anosov system, and that the KS entropy of the system $h_{\mathrm{KS}} = h$, the entropy obtained from (9.4), if the starting partition was a Markov partition.

Markov partitions have other useful properties as well. They allow one to represent the dynamics of the system as a Markov process. That is, the nice stretching and contracting properties of the sides of the rectangles of a Markov partition allow us to define a Markov transition matrix for the dynamics of a probability distribution function for the system which is coarse grained on the rectangles. By taking finer and finer rectangles in the Markov partition, one can represent the dynamics of the system to any required precision, and the stationary state probabilities will define an approximation to the equilibrium state of the system. One further property of Markov partitions is that in simple cases, at least, they can be used to prove a very important dynamical theorem known as *Pesin's theorem*, which relates the KS entropy of a closed, Anosov system to the sum of the positive Lyapunov exponents for the system, provided that the Lyapunov exponents are calculated as ensemble averages with respect to the ergodic measure of the system, or equivalently, as time averages, starting from a typical point, not a point on a periodic orbit, say, of the system:

(Pesin's Theorem) *For a closed Anosov system (a system is closed when phase space is mapped onto itself), with an invariant, Sinai–Ruelle–Bowen measure, the Kolmogorov–Sinai entropy is equal to the sum of the positive Lyapunov exponents,*

$$h_{\mathrm{KS}} = \sum_i \lambda_+^i. \tag{9.8}$$

where the KS entropy and the Lyapunov exponents are calculated using the SRB measure.

We have not yet discussed Sinai–Ruelle–Bowen (SRB) measures, but they will be the subject of Chapter 13. For the moment, it is sufficient to say that the measures we have used to compute the KS entropy and Lyapunov exponents in the examples given above are SRB measures. The central feature of SRB measures needed to prove Pesin's theorem is that an SRB measure is smooth along expanding directions, as are the measures we have studied so far. In the case that the measure used to compute the KS entropy and the Lyapunov exponents is not an SRB measure, one has Ruelle's inequality, which states that the KS entropy is always less than or equal to the sum of the positive Lyapunov exponents, i.e., $\sum_i \lambda_+^i \geq h_{\mathrm{KS}}$.

Thus from Pesin's theorem, we should not be surprised that the KS entropy of the baker's map is $\ln 2$ or that the KS entropy of the Arnold cat map is $\ln[(3+\sqrt{5})/2]$.

We will return to this theorem when we consider open systems for the escape-rate formalism in Chapter 11. There we will see an important application of Ruelle's inequality and breakdown of Pesin's theorem when the appropriate measure is not smooth along the expanding directions.

9.4 Further reading

Extensive discussions of the Kolmogorov–Sinai entropy, Markov partitions, hyperbolic, Anosov, and Axiom-A systems can be found in Katok and Hasselblatt [KH95], Sinai [Sin89], Bowen [Bow78b], Ruelle [Rue89], as well as in Ott [Ott92], and Toda *et al.* [TKS92]. The definition of the KS entropy using the notion of separated sets can be found in the paper by Katok [Kat82]. The complicated structure of Markov partitions are described in papers by Bowen [Bow78a]and by Cawley [Caw91]. Pesin's theorem is discussed in the book by Pollicott [Pol93] and a key paper is by Ledrappier and Young [LY85]. A nice proof of Pesin's theorem for two-dimensional hyperbolic toral automorphisms is due to Berg and is discussed in the book of Peterson [Pet91].

9.5 Exercises

1. **A Markov partition for the baker's map.** Consider a simple Markov partition for the baker's map obtained by dividing the map into quarters by putting lines at $x = 1/2$ and $y = 1/2$. Use the Frobenius–Perron equation for this map to derive a set of Markovian equations for the time dependence of the probabilities that the system will be found in each of the quarters, starting from some initial set of probabilities. What is the stationary set of probabilities for each quarter and the probability density in each quarter?

2. Carry out a similar calculation using the transformation given by (13.1) using a partition of the unit square into four rectangles obtained by drawing a horizontal line at $y = r$ and a vertical line

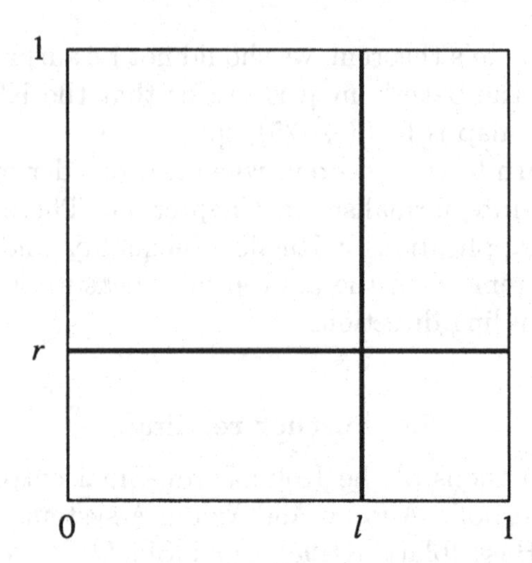

Fig. 9.3 The unit square partitioned into four rectangles by a horizontal line at $y = r$ and a vertical line at $x = l$.

at $x = l$ (Fig. 9.3). What are the stationary probabilities and probability densities for this model?

10
The Frobenius–Perron equation

In our discussions of the baker's transformation, we had occasion to use a discrete-time form of Liouville's equation at a number of points without calling any particular attention to the fact that we were doing so. Now it is time to look more carefully at Liouville's equation for discrete maps and to explore some of the interesting properties of this equation, which is referred to in the literature as the Frobenius–Perron equation.

Suppose we have some map $M(\Gamma)$ of the phase space onto, or perhaps into, itself.† Let us also suppose that there is some distribution of points on the phase space which is characterized by a distribution function at some initial time $t = 0$, $\rho_0(\Gamma)$. Then after one time step, the distribution changes because the points move on the phase-space according to the mapping M. We denote the distribution function after the nth application of the map by $\rho_n(\Gamma)$. Then $\rho_n(\Gamma)$ satisfies the equation

$$\rho_n(\Gamma) = \int d\Gamma' \delta(\Gamma - M(\Gamma'))\rho_{n-1}(\Gamma'), \tag{10.1}$$

which is clear once one realizes that the delta function only registers at points Γ' that are mapped to the point Γ by M. This is the Frobenius–Perron equation. One might note that a slightly different form of the Frobenius–Perron equation can be given by evaluating the delta function at the zeroes of its argument. Then one gets a relation between the distribution functions at sequential time steps in terms of the derivative, or Jacobian derivative, of the map M. We will see examples of this below.

One immediate application of this equation occurs for the case where a phase space is closed and mapped onto itself by M. Then

† We can include the possibility that some points of phase space may be mapped outside of the phase space. We will see an example of this when we consider open systems in the next chapter.

one can ask if there is a steady state or an equilibrium distribution function for such a system. If the system is isolated, then the map may lead to an equilibrium state for which the distribution function $\rho_{\text{eq}}(\Gamma)$ satisfies

$$\rho_{\text{eq}}(\Gamma) = \int d\Gamma' \delta(\Gamma - M(\Gamma'))\rho_{\text{eq}}(\Gamma'), \tag{10.2}$$

with the same distribution inside the integral as on the left-hand side. If the map M is one-to-one and onto i.e., a *bijection* in the mathematical terminology) then it is easy to see that $\rho_{\text{eq}}(\Gamma)$ could be any function of the globally conserved quantities, such as the energy and number of particles in typical applications to statistical mechanics, and satisfy (10.2).

Another application of the Frobenius–Perron equation was used tacitly when we discussed periodic points under the baker's transformation in the unit square. In such a case, one can define delta function distribution on the periodic points. Suppose we have identified a period-p sequence of points in the unit square such that $\Gamma_0 \to \Gamma_1 \to \Gamma_2 \to \cdots \to \Gamma_{p-1} \to \Gamma_0$. Then an initial distribution $\rho_0 = \delta(\Gamma - \Gamma_0)$ will be transformed as $\rho_0 \to \rho_1 = \delta(\Gamma - \Gamma_1) \to \cdots \rho_{p-1} = \delta(\Gamma - \Gamma_{p-1}) \to \rho_0$, according to the Frobenius–Perron equation. Thus a distribution function defined on the periodic points is periodic, too.

10.1 One-dimensional systems

Now we turn to another example of (10.2) that occurs in the theory of Lyapunov exponents. Consider a one-dimensional system on some interval I, and a map of this interval onto itself, $M : x_n \to x_{n+1} = M(x_n)$ (see Fig. 10.1). Let's consider an infinitesimal region dx_0 about the point x_0 on the interval. In order to determine the Lyapunov exponent associated with this map at the point x_0 we look at the growth of the initial interval dx_0 over t iterations of the map. We suppose that this region is still infinitesimally small, and apply the chain rule:

$$\left|\frac{dx_t}{dx_0}\right| = \left|\frac{dx_t}{dx_{t-1}}\right| \left|\frac{dx_{t-1}}{dx_{t-2}}\right| \cdots \left|\frac{dx_1}{dx_0}\right|.$$

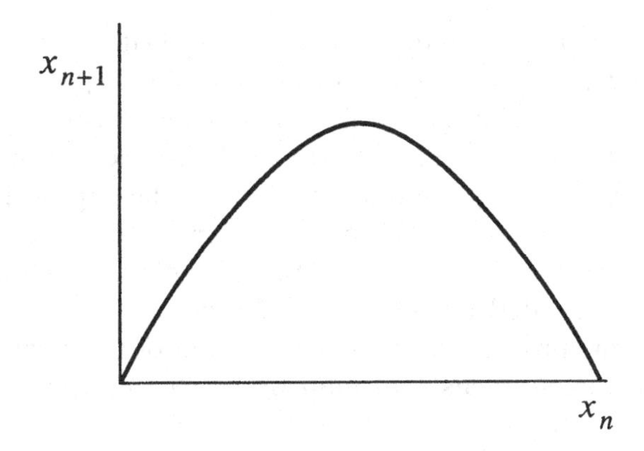

Fig. 10.1. A one-dimensional map.

We now use the fact that the derivatives here are simply the derivatives of the map function M

$$\left|\frac{dx_j}{dx_{j-1}}\right| = |M'(x_{j-1})| \,.$$

Then

$$\left|\frac{dx_t}{dx_0}\right| = |M'(x_{t-1})|\,|M'(x_{t-2})|\cdots|M'(x_0)| \,. \tag{10.3}$$

Suppose that the growth of the initial region is exponential, with a local Lyapunov exponent $\lambda(x_0)$, i.e.,

$$\left|\frac{dx_t}{dx_0}\right| \propto e^{\lambda(x_0)t}. \tag{10.4}$$

Then

$$\lambda(x_0) = \frac{1}{t}\sum_{j=0}^{t-1} \ln|M'(x_j)| \,. \tag{10.5}$$

In the limit $t \to \infty$,

$$\lambda(x_0) \to \overline{\ln|M'(x)|}\Big|_{x=x_0} , \tag{10.6}$$

where the overbar denotes the long-time average of $\ln|M'|$.

If the map is ergodic then the time average is equal to the ensemble average. In that case λ is a constant, and is given by

$$\lambda = \int d\mu(x) \ln|M'(x)| = \int dx \rho_{\text{eq}}(x) \ln|M'(x)|. \tag{10.7}$$

In the second equality in (10.7), we have assumed that the measure element $d\mu$ can be expressed in terms of a phase-space density as $d\mu = \rho_{\text{eq}}(x)dx$. Measures for which this is true are called *absolutely continuous measures* with respect to the Lebesgue measure. There are cases, often involving fractal sets, where the appropriate measure is not absolutely continuous with respect to the appropriate Lebesgue measure. We will encounter examples of such measures when we discuss SRB measures in Chapter 13. .

We can now proceed with the calculation of the Lyapunov exponent for a one-dimensional map by using the property of the delta function that

$$\delta(x - M(y)) = \sum_i \frac{1}{|M'(y_i)|}\delta(y - y_i), \tag{10.8}$$

where y_i are the solutions of $M(y) = x$. Then, in general, we have

$$\rho_t(x) = \sum_i \frac{1}{|M'(y_i)|}\rho_{t-1}(y_i). \tag{10.9}$$

If the density is the equilibrium density, we have

$$\rho_{\text{eq}}(x) = \sum_i \frac{1}{|M'(y_i)|}\rho_{\text{eq}}(y_i). \tag{10.10}$$

As as application of this construction, consider the following map on the unit interval:

$$x_{n+1} = \begin{cases} x_n/a & \text{if } 0 < x_n < a, \\ (1 - x_n)/b & \text{if } 1 \geq x_n > a, \end{cases} \tag{10.11}$$

with $a + b = 1$ (see Fig. 10.2). Equation (10.9) gives

$$\rho_t(x) = a\rho_{t-1}(y_1) + b\rho_{t-1}(y_2), \tag{10.12}$$

and an equilibrium distribution must satisfy

$$\rho_{\text{eq}}(x) = a\rho_{\text{eq}}(y_1) + b\rho_{\text{eq}}(y_2). \tag{10.13}$$

One can easily check that a normalized equilibrium solution is $\rho_{\text{eq}} = 1$ since $a + b = 1$. Equation 10.7 for the Lyapunov exponent can be used to obtain

$$\lambda = \int_0^1 dx\rho_{\text{eq}}(x)\ln|M'(x)| = a\ln\left(\frac{1}{a}\right) + b\ln\left(\frac{1}{b}\right) > 0. \tag{10.14}$$

Next we calculate the KS entropy, (9.3), of this map as an example of Pesin's theorem. We start with a partition of the unit interval into intervals of lengths a and b. Then $H(1) = a\ln(1/a) + b\ln(1/b)$. The pre-image of the set of length a is composed of two

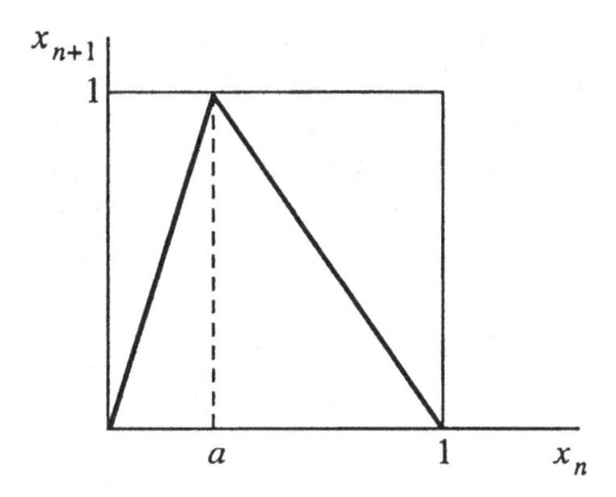

Fig. 10.2. A simple map of the unit interval onto itself.

sets, one of length a^2 and one of length ab. Similarly, the preimages of the set of length b are sets of lengths b^2 and ba (see Fig. 10.2). So

$$\begin{aligned} H(2) &= a^2 \ln\left(\frac{1}{a^2}\right) + ab \ln\left(\frac{1}{ab}\right) + b^2 \ln\left(\frac{1}{b^2}\right) + ba \ln\left(\frac{1}{ba}\right) \\ &= 2a^2 \ln\left(\frac{1}{a}\right) + 2ab \ln\left(\frac{1}{a}\right) + 2ab \ln\left(\frac{1}{b}\right) + 2b^2 \ln\left(\frac{1}{b}\right) \\ &= 2a(a+b) \ln\left(\frac{1}{a}\right) + 2b(b+a) \ln\left(\frac{1}{b}\right) \\ &= 2H(1). \end{aligned} \tag{10.15}$$

In general, $H(n) = nH(1)$, so that

$$h_{\mathrm{KS}} = \lim_{n\to\infty} \frac{1}{n} H(n) = H(1) \tag{10.16}$$

or

$$h_{\mathrm{KS}} = a \ln\left(\frac{1}{a}\right) + b \ln\left(\frac{1}{b}\right) = \lambda, \tag{10.17}$$

as expected.

10.2 The Frobenius–Perron equation in higher dimensions

Now let us consider the Frobenius–Perron equation in a higher-dimensional setting. Returning to (10.2), we note that the delta function in the integrand is a product of several delta functions, one for each dimension of the phase-space. Also, we have to evaluate these delta functions at the zeroes of their arguments. Depending upon the map M there may be none, one, or many values of Γ' for each point Γ. At any rate, we shall assume that there is at least one solution for Γ' for a given phase point Γ. We then expand the arguments of the delta function about each of the solution points in a Taylor series of the general form

$$\begin{aligned} x - f(x', y', z', \ldots) &= x - f(x'_0, y'_0, z'_0, \ldots) \\ &\quad - (x' - x'_0) f_{x'}(x'_0, y'_0, z'_0, \ldots) \\ &\quad - (y' - y'_0) f_{y'}(x'_0, y'_0, z'_0, \ldots) \\ &\quad + \cdots, \end{aligned} \tag{10.18}$$

where $f_{x'}$ denotes a partial derivative of f with respect to the x' variable, and $(x'_0, y'_0, z'_0, \ldots)$ denotes a point in phase-space where $x = f(x'_0, y'_0, z'_0, \ldots)$. If we make a similar Taylor expansion in the argument of each delta function, then (10.2) can be written in the form

$$\rho_n(\Gamma) = \sum_{\Gamma_i} \frac{1}{|J(\Gamma_i)|} \rho_{n-1}(\Gamma_i), \tag{10.19}$$

where the sum is over all the pre-images Γ_i of the point Γ. Here, $J(\Gamma_i)$ is the Jacobian of the transformation $M(\Gamma)$,

$$J(\Gamma_i) = \frac{\partial M(\Gamma)}{\partial \Gamma}, \tag{10.20}$$

evaluated at $\Gamma = \Gamma_i$.

For a measure-preserving transformation, the Jacobian is unity, and if the transformation is one-to-one (bijective), (10.19) reduces to a discrete-time version of the Liouville equation studied earlier for Hamiltonian mechanical systems. In Exercise 7.3, you encountered the Frobenius–Perron equation for the baker's transformation and were asked to show that it is identical to the Liouville equation which we obtained for this map. In Chapter 15, we will consider properties of a Frobenius–Perron operator which is obtained by regarding the right-hand side of the Frobenius–Perron equation as an integral operator acting on a distribution function.

10.3 Further reading

Detailed discussions of the Frobenius–Perron equation can be found in the books by Lasota and Mackey [LM94] and by Ott [Ott92].

10.4 Exercises

1. Derive the Frobenius–Perron equation for the dyadic map on the unit interval in one dimension, $M(x) = 2x \pmod 1$.

2. Derive the Frobenius–Perron equation for the following map on the unit square

$$M(x,y) = \begin{cases} (x/l, ry) & \text{for } 0 \leq x < l, \\ ((x-l)/r, r+ly) & \text{for } l \leq x \leq 1. \end{cases} \tag{10.21}$$

where $l + r = 1$, and $l, r > 0$.

11
Open systems and escape rates

11.1 The escape-rate formalism

One of the most interesting developments in the theory of irreversible processes is a connection, first explored by Gaspard and Nicolis, between the dynamical properties of *open* systems and the hydrodynamic or transport properties of such systems. We will explore this connection in Chapter 13, but first we must consider the dynamical properties of open systems. We consider a system to be open if the phase space of the system has physical boundaries and there is a mapping or transformation which can take phase points inside the boundaries to phase points outside the boundaries. Further, we will assume that the boundaries are such that once a phase point passes a boundary, it can never return to the bounded system. Thus the boundaries on the phase space region may be considered to be absorbing. To get some idea of the motivation for considering open systems, we might imagine a Brownian particle diffusing in a fluid inside a container with absorbing boundaries. The motion of the particle is really deterministic and can be described – microscopically – by some transformation in the phase space of the entire system. Now, when we describe the motion of the phase point, we lose the Brownian particle whenever it encounters the boundary of the container. If we were to describe the motion macroscopically, we would solve the diffusion equation for the probability density of the Brownian particle, in the fluid, with absorbing boundaries. The probability of finding the particle inside the container is an exponentially decreasing function of time with decay coefficient depending on the diffusion coefficient of the Brownian particle in the fluid, and on the geometry of the container. The escape-rate method, which we are about to develop, relates this decay rate to dynamical properties of the deterministic microscopic dynamics of the system. We note immediately that Pesin's theorem does not apply to open systems. In fact, the difference between the sum of the positive Lyapunov exponents and the KS entropy is going to be an important quantity in our discussions. We remind the reader that Pesin's theorem shows that this difference vanishes for closed systems that have an invariant SRB measure, smooth in unstable directions.

This subject is sufficiently complicated that it is best to consider an explicit example before making any general statements. As a working example, we take the following discrete map with a possibility for points to escape:

$$x_{n+1} = \begin{cases} x_n/p_0 & \text{for } 0 < x_n < \frac{1}{2}, \\ [(x_n - 1)/p_1] + 1 & \text{for } \frac{1}{2} < x_n < 1, \end{cases} \tag{11.1}$$

where $p_0, p_1 < \frac{1}{2}$, and $p_0 + p_1 < 1$ (see Fig. 11.1).

At each step of the mapping, points between p_0 and $1-p_1$ leave the unit interval. Let's start the system with points uniformly distributed on the unit interval $0 \leq x \leq 1$. The points that remain inside the unit interval after one application of the mapping form two sets of lengths p_0 and p_1, respectively. Initial points that remain inside the unit interval after two mappings fall into four sets of lengths p_0^2, p_0p_1, p_1p_0, and p_1^2, respectively (see Fig. 11.1). Iterating this map infinitely many times, we see that the set of initial points which never leave the unit interval form a very special set whose properties we are about to determine. It will be clear shortly that this set of initial points is in fact a Cantor set.

We begin by determining the total length of the points that remain in the unit interval after n steps. We see from the above argument that the total measure (in this case, the length of the two intervals) of the points that survive one iteration is $l_1 = (p_0 + p_1)$. The total measure of the initial points that survive two iterations of the map is $l_2 = (p_0+p_1)^2$ and consists of four intervals. The total length of the intervals containing points that survive n iterations decays exponentially: $l_n = (p_0 + p_1)^n \to 0$ as $n \to \infty$. Therefore, after n steps, the density of points in the unit interval is decreasing as l_n and thus decaying exponentially,

$$l_n = e^{n \ln(p_0+p_1)} \equiv e^{-n\gamma},$$

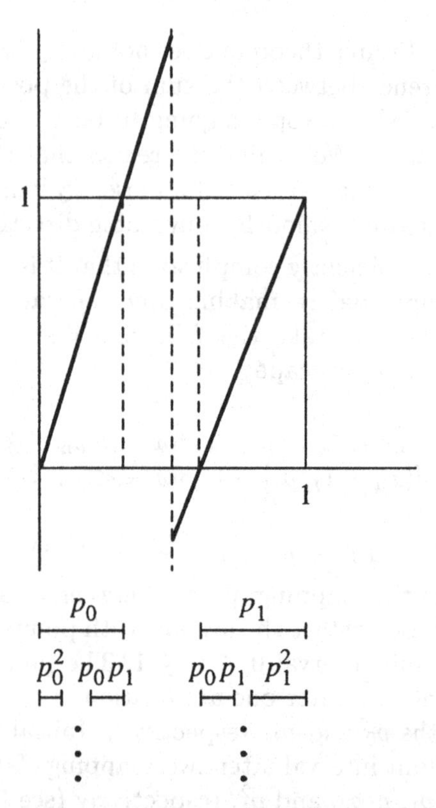

Fig. 11.1 A linear map with escape from the unit interval. The first two steps in the construction of the fractal repeller are illustrated.

where γ is the escape rate, given by

$$\gamma = \ln\left(\frac{1}{p_0 + p_1}\right). \tag{11.2}$$

Although the measure of the set of initial points which never leave the unit interval is zero, this set does contain a non-countable number of points. To see this, note that each of the never-escaping points discussed above can be coded by sequences of zeros and ones, according to which side of the line $x = \frac{1}{2}$ the point finds itself at each iteration. We have already seen in our discussions of the binary representations of numbers in connection with the baker's transformation that these sequences of zeros and ones stand in a one-to-one relation to the points in the interval from 0 to 1. Thus the set of points that are trapped forever in the unit interval under the mapping is a set of zero Lebesgue measure with an uncountable infinity of points. This is a Cantor set. In fact, if $p_0 = p_1 = \frac{1}{3}$, the set which we have just constructed is the *middle-third* Cantor set.

The set of points that remain forever is called a *repeller*, and denoted by F_{R}, because the other points leave the unit interval sooner or later no matter how close they are to points of the repeller initially. We are now going to show that we can define Lyapunov exponents and a KS entropy for the repeller, and, for the case of repeller, the sum over the positive Lyapunov exponents and the Kolmogorov–Sinai entropy are not equal. For the example we are discussing, there is one positive Lyapunov exponent, $\lambda(F_{\mathrm{R}})$, and the difference between it and the KS entropy, $h_{\mathrm{KS}}(F_{\mathrm{R}})$, will be shown below to be the escape rate,

$$\lambda(F_{\mathrm{R}}) - h_{\mathrm{KS}}(F_{\mathrm{R}}) = \gamma.$$

To do this, we first consider the set of initial points that survive n iterations and we define a normalized measure on these 2^n intervals. Each of these intervals has a length l_i with total length $\sum_i l_i = (p_0+p_1)^n$. The normalized measure of an interval of length l_i is then taken to be

$$\mu_i = \frac{l_i}{(p_0 + p_1)^n}.$$

Then $\sum_{i=1}^{2^n} \mu_i = 1$. To define a Lyapunov exponent on the repeller, we consider two points which are both in one of these 2^n intervals. These two points will be characterized by the same sequence of travels from one side of $x = \frac{1}{2}$ to the other over the n steps, but their separation will increase at each step. If the two points are infinitesimally close initially, their separation will be, after n steps,

$$\begin{aligned}\left|\frac{dx_n}{dx_0}\right| &= M'(x_{n-1})M'(x_{n-2})\cdots M'(x_0) \\ &= \left(\frac{1}{p_0}\right)^{n_0}\left(\frac{1}{p_1}\right)^{n_1},\end{aligned}$$

where n_0 (n_1) is the number of times that the points are interval are to the left (right) of the line $x = \frac{1}{2}$ over the n steps, so that

$n_0 + n_1 = n$. Therefore,

$$\left|\frac{dx_n}{dx_0}\right| = \exp[n_0 \ln(1/p_0) + n_1 \ln(1/p_1)] = e^{n\lambda(x_0)},$$

where, for large n, $\lambda(x_0)$ is the local Lyapunov exponent. Then,

$$\lambda(x_0) = \frac{n_0}{n} \ln\left(\frac{1}{p_0}\right) + \frac{n_1}{n} \ln\left(\frac{1}{p_1}\right).$$

Suppose that the map is ergodic when restricted to points on the repeller. If so, then we would expect that the fraction of the time that a point on the repeller falls to the left of $x = \frac{1}{2}$ should be equal to the fraction of the set of initial points that is to the left of $x = \frac{1}{2}$, namely,

$$\begin{aligned} \frac{n_0}{n} &= \frac{p_0}{p_0 + p_1}, \\ \frac{n_1}{n} &= \frac{p_1}{p_0 + p_1}. \end{aligned} \tag{11.3}$$

Then for a typical point, the Lyapunov exponent would be

$$\lambda = \frac{1}{p_0 + p_1}\left[p_0 \ln\left(\frac{1}{p_0}\right) + p_1 \ln\left(\frac{1}{p_1}\right)\right]. \tag{11.4}$$

We will make this calculation more precise in a moment, but it is correct. Now we use similar arguments to compute the KS entropy.

In order to calculate h_{KS}, we use the usual definition in terms of the normalized measure and consider the set of points that survive n iterations of the map. Using the normalized measure defined above, we take the entropy of this set to be

$$H(n) = \sum_{i=1}^{2^n} \mu_i \ln\left(\frac{1}{\mu_i}\right), \tag{11.5}$$

with

$$\begin{aligned} \mu_i &= \frac{p_0^{n_0} p_1^{n_1}}{(p_0 + p_1)^n}, \\ n &= n_0 + n_1. \end{aligned} \tag{11.6}$$

Then

$$\begin{aligned} H(n) &= -\sum_i^{2^n} \frac{p_0^{n_0} p_1^{n_1}}{(p_0 + p_1)^n} \ln\left[\frac{p_0^{n_0} p_1^{n_1}}{(p_0 + p_1)^n}\right] \\ &= -\sum_{n_0=0}^{n} \frac{n!}{n_0!(n - n_0)!} \frac{p_0^{n_0} p_1^{n-n_0}}{(p_0 + p_1)^n} \ln\left[\frac{p_0^{n_0} p_1^{n-n_0}}{(p_0 + p_1)^n}\right], \end{aligned} \tag{11.7}$$

where we have inserted the combinatorial factor, which tells us how many different intervals have the same values of n_0 and n_1. We compute $H(n)$ as

$$\begin{aligned} H(n) &= \frac{n\ln(p_0+p_1)}{(p_0+p_1)^n}\sum_{n_0=0}^{n}\frac{n!}{n_0!(n-n_0)!}p_0^{n_0}p_1^{n-n_0} \\ &- \sum_{n_0=0}^{n}\frac{n!}{n_0!(n-n_0)!}\frac{p_0^{n_0}p_1^{n-n_0}}{(p_0+p_1)^n}\times \\ &\qquad [n_0\ln(p_0)+(n-n_0)\ln(p_1)]. \end{aligned} \tag{11.8}$$

In the first term, Newton's binomial expansion can be used, and then the factor $(p_0+p_1)^n$ cancels. If this binomium is differentiated with respect to p_0 or p_1, the resulting formula can be used for the second and third terms. Then we obtain

$$H(n) = n\ln(p_0+p_1)+\frac{np_0}{p_0+p_1}\ln\left(\frac{1}{p_0}\right)+\frac{np_1}{p_0+p_1}\ln\left(\frac{1}{p_1}\right). \tag{11.9}$$

Now, if we define $h_{\rm KS}$ in the usual way using (9.4),

$$h_{\rm KS}(F_{\rm R}) = \lim_{n\to\infty}\frac{1}{n}H(n), \tag{11.10}$$

we find, using (11.4), that

$$h_{\rm KS}(F_{\rm R}) = \ln(p_0+p_1)+\lambda. \tag{11.11}$$

The first term on the right-hand side shows that for an open system, $h_{\rm KS}\neq\lambda$. The measure on the repeller is not smooth in the expanding direction, since it is not absolutely continuous with respect to the Lebesgue measure in the x-direction. However, if $p_0+p_1 = 1$, then there wouldn't be any escape possibility, the measure of the map would be smooth in the x-direction, so the measure would be an SRB measure, and Pesin's theorem would apply. Indeed, then $h_{\rm KS}=\lambda$. Using the expression for the escape rate, (11.2), we conclude that for this map

$$\gamma = \lambda(F_{\rm R}) - h_{\rm KS}(F_{\rm R}). \tag{11.12}$$

This is the escape-rate formula which we will make use of in formulating a new expression for the coefficient of diffusion in Chapter 13. The escape-rate formalism was developed by Kantz and Grassberger, by Bohr and Rand, by Tel, and by Kadanoff and Tang. The reader may wish to write a simple program to determine the exit time for points on the unit interval. This exit time for points turns out to be a wild function defined almost everywhere on the interval $0 \leq x \leq 1$.

11.2 The Smale horseshoe

One of the central constructions of dynamical systems theory is the Smale horseshoe. It was invented by Smale to characterize the complicated dynamics on infinitely fine scales first noticed by Poincaré, which are often called *homoclinic* or *heteroclinic tangles.* That is, if one can detect the presence of a Smale horseshoe in a dynamical system, then it follows that there will be such a tangle in the system's dynamics. We refer the reader to the standard books on chaos theory for a discussion of this aspect of the horseshoe. Here, we will use the horseshoe to illustrate some of the properties of a fractal repeller for some two-dimensional, measure-preserving maps.

Consider a unit square, Fig. 11.2 (a), and a measure-preserving transformation H of the square that stretches sets along the x-direction uniformly by a factor of η, with $\eta \gg 2$, and compresses sets along the y-direction uniformly by a factor of $1/\eta$. Now take this long, thin rectangle and fold it into a horseshoe as illustrated in Fig. 11.2 (b). The intersection of the folded horseshoe with the original square is the set of initial points which remain in the square after one iteration. The area or measure of this set is denoted by $\mu_+(1) = 2/\eta$. If we iterate the map again, Fig 11.2 (c), we get a set of points (indicated by the horizontal hatched strips) in the expanding direction of the map, with total measure $\mu_+(2) = (2/\eta)^2$, which are the sets of initial points that remain in the square after two iterations of the map. Clearly, after n iterations we get a set of 2^n horizontal lines with total measure $\mu_+(n) = (2/\eta)^n$, representing the set of initial points that remain after n iterations. As n gets very large, we get a collection of sets oriented along the unstable direction, whose cross-section in the y-, or stable, direction is approaching a Cantor set. In view of the discussion in the previous section, this horizontal Cantor set will be called the *forward repeller.* The escape rate γ is obviously $\gamma = \ln \eta - \ln 2$.

For an invertible, measure-preserving system, we would like to define a repeller as a set that is invariant under both the map

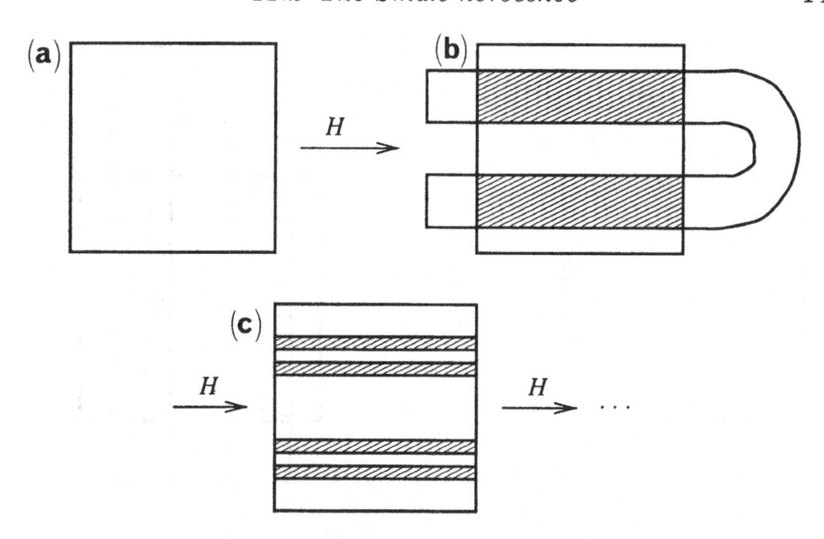

Fig. 11.2. Illustration of the Smale horseshoe transformation.

and its inverse. The forward repeller is not an invariant set under H^{-1}. However, we can, as illustrated in Fig. 11.3, construct a time-reversed repeller by carrying out the above procedure for H^{-1}. This leads to a vertical Cantor set, aligned along the stable direction. The escape rate is identical to that of the forward map.

The actual invariant repeller is constructed by taking the intersection of the forward- and time-reversed repellers, as illustrated in Fig. 11.4. As $n \to \infty$, we obtain a *Cantor dust* in the unit square, which is the set of initial points that never escape in either the forward- or time-reversed motions. It is left as a problem to show that:

1. The positive Lyapunov exponent on the repeller is $\ln \eta$;
2. The KS entropy on the repeller is $\ln 2$; and
3. The points on the repeller can be labeled by infinite sequences of zeros and ones just as one does with points in the unit square for the baker's map. The dynamics on the repeller can then be mapped onto the same Bernoulli shift as in the baker's map.

Here we see that, for an invertible measure-preserving transformation, we can define the repeller as an *invariant* set in phase space, with a dynamics defined on it and with dynamical quantities that satisfy the escape-rate formula. The Lebesgue measure of

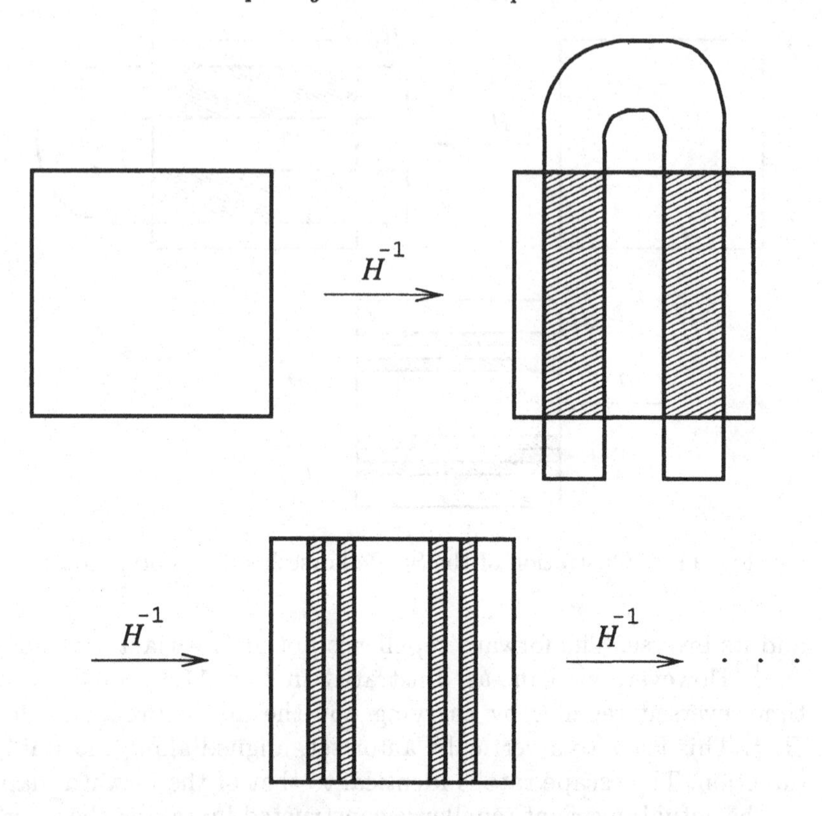

Fig. 11.3. Time-reversed Smale horseshoe transformation.

the repeller is zero, of course, but this is only one way to characterize the 'size' of a set. A more sensitive way to characterize the size of a set is the *box-counting dimension*, which we now discuss.

11.3 The box-counting dimension

In our first example of a repeller, we constructed a Cantor set on the unit interval with zero Lebesgue measure. However, this Cantor set of initial points that stay forever in the unit interval can still be described as having a dimension. In fact, a non-integer dimension can be ascribed to this set, called the box-counting dimension. Sets with non-integer dimensions are called *fractals*. To construct the box-counting dimension of a set, we begin by

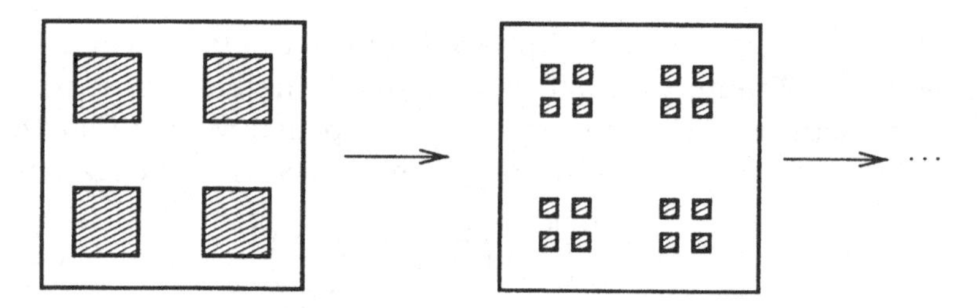

Fig. 11.4 Illustration of the Smale horseshoe's invariant repeller.

covering the set entirely with 'cubical' boxes of length ϵ on a side. The dimensions of the 'cubes' can be arbitrarily large, if needed. Suppose $N(\epsilon)$ boxes are needed to cover the set of points. For small ϵ, $N(\epsilon)$ may have to be large. We define the box-counting dimension, d, of the set of interest in terms of the limit of $N(\epsilon)$ as ϵ tends to zero:

$$N(\epsilon) = \left(\frac{1}{\epsilon}\right)^d, \tag{11.13}$$

or

$$d = \lim_{\epsilon \to 0} \frac{\ln[N(\epsilon)]}{\ln(1/\epsilon)}. \tag{11.14}$$

To test this formula, consider an ordinary line of length L. Then one easily sees that $N(\epsilon) \propto L/\epsilon$, so that the box-counting dimension d is just 1.

Let's apply this method to find the box-counting dimension of the repeller. As in the computations of λ and h_{KS}, we consider first the set of points that survive n iterations of the map. This set consists of 2^n intervals with lengths p_0^n, $p_0^{n-1}p_1$, ... , p_1^n. Now the points that survive $n+1$ iterations consist of 2^{n+1} intervals which are similar to the intervals listed above, except that 2^n have the lengths listed above but scaled by a factor of p_0, and the other 2^n intervals are smaller by a factor of p_1. Now let $N_n(\epsilon)$ be the number of intervals of length ϵ needed to cover the 2^n intervals that survive n iterations. By the above scaling argument, it follows that $N_{n+1}(\epsilon)$ must satisfy

$$N_{n+1}(\epsilon) = N_n\left(\frac{\epsilon}{p_0}\right) + N_n\left(\frac{\epsilon}{p_1}\right), \tag{11.15}$$

since one will need intervals of length ϵ/p_0 to cover one set of 2^n

intervals at the nth step, and intervals of length ϵ/p_1 to cover the other set of 2^n intervals. When we look at the limit $n \to \infty$ in order to determine the properties of the repeller, we find that the box-counting dimension is given by the transcendental equation

$$N(\epsilon) = \left(\frac{1}{\epsilon}\right)^d = \left(\frac{p_0}{\epsilon}\right)^d + \left(\frac{p_1}{\epsilon}\right)^d, \tag{11.16}$$

or, eliminating the parameter ϵ,

$$1 = p_0^d + p_1^d. \tag{11.17}$$

One can easily see that if $p_0 + p_1 < 1$, then d will not be an integer. The arguments that we have just used is based upon the *self-similarity* property of the fractal repeller for our model. That is, the fractal set looks the same at every level of magnification.

One of the most famous fractal sets is Cantor's middle-third set. As we mentioned earlier, this Cantor set appears as a repeller if we take $p_0 = p_1 = \frac{1}{3}$. Then the box-counting dimension is obtained from the relation $1 = 2(\frac{1}{3})^d$. Thus for the middle-third set, the box-counting dimension is

$$d = \frac{\ln 2}{\ln 3}. \tag{11.18}$$

Therefore, we can argue on the basis of this simple example that in the general case, a repeller will be a fractal object in phase space with a non-integer box-counting dimension. The box-counting dimension is just one of a set of possible dimensions that may be constructed for sets. In many ways a more useful dimension is the Hausdorff dimension, d_{H}. The Hausdorff dimension of a set, A, is obtained by covering the set with a finite or countable number of open sets with diameter less than or equal to ϵ. Then define a function $m_{\mathrm{H}}(A, \alpha)$ by

$$m_{\mathrm{H}}(A, \alpha) = \liminf_{\epsilon \to 0} \inf_{\mathcal{G}} \left[\sum_{U \in \mathcal{G}} (\operatorname{diam} U)^\alpha \right], \tag{11.19}$$

where $\mathcal{G}$ denotes a finite or countable covering of A by open sets, U, of diameter $\leq \epsilon$. We note that it is possible for $m_{\mathrm{H}}(A, \alpha)$ to be infinite. The Hausdorff dimension, $d_{\mathrm{H}}(A)$ of the set A is then defined as

$$d_{\mathrm{H}}(A) = \inf\{\alpha : m_{\mathrm{H}}(A, \alpha) = 0\}. \tag{11.20}$$

For the middle third Cantor set we can easily show that

$$m_{\mathrm{H}}(\alpha) = \lim_{n\to\infty} \exp[n(\ln 2 - \alpha \ln 3)], \tag{11.21}$$

so that $d_{\mathrm{H}} = \ln 2/\ln 3$. Note that for α less than this value $m_{\mathrm{H}}(\alpha)$ is infinite, so that the Hausdorff dimension locates the transition of m_{H} from infinity to zero. The theory behind this definition of dimension, as well as other dimensions that can be usefully employed to characterize fractal sets, is discussed in the books by Falconer and by Pesin (see *Further reading*).

11.4 The Lyapunov exponent for the repeller

We can now return to the calculation of the Lyapunov exponent for the repeller, $\lambda(F_{\mathrm{R}})$, and re-examine the argument used to obtain (11.4). Again, let's look at the 2^n intervals that survive the first n iterations. Each of these intervals has a length $l_i = p_0^{n_i} p_1^{n-n_i}$, a natural measure $\mu_i = l_i/(p_0 + p_1)^n$, and a 'stretching factor' Λ_i for two nearby points in the interval, defined by

$$\Lambda_i = \left|\frac{dx_n}{dx_0}\right| = p_0^{-n_i} p_1^{-(n-n_i)} = l_i^{-1}. \tag{11.22}$$

Here, n_i is simply the number of times a point in this interval finds itself to the left of $x = \frac{1}{2}$ in the course of the n iterations. There may be several intervals with the same value for n_i, of course. If we now examine the definition of the Lyapunov exponent $\lambda(F_{\mathrm{R}})$ for the repeller, we see that it can be written in terms of the average (using the natural measure on the repeller) of the logarithms of the stretching factors as

$$\lambda(F_{\mathrm{R}}) = \lim_{n\to\infty} \frac{1}{n} \sum_{i=1}^{2^n} \mu_i \ln \Lambda_i = - \lim_{n\to\infty} \frac{1}{n} \sum_{i=1}^{2^n} \mu_i \ln l_i. \tag{11.23}$$

When the various quantities are inserted into (11.23), we find that

$$\begin{aligned} \lambda(F_{\mathrm{R}}) &= \lim_{n\to\infty} \frac{1}{n} \frac{1}{(p_0 + p_1)^n} \sum_i l_i \ln\left(\frac{1}{l_i}\right) \\ &= \lim_{n\to\infty} \frac{1}{n} \frac{1}{(p_0 + p_1)^n} \sum_{n'=1}^{n} \frac{n!}{n'!(n-n')!} \\ &\quad \times p_0^{n'} p_1^{n-n'} \ln\left[\frac{1}{p_0^{n'} p_1^{n-n'}}\right]. \end{aligned} \tag{11.24}$$

This leads to our previous result, (11.4),

$$\lambda = \frac{1}{p_0 + p_1}\left[p_0 \ln\left(\frac{1}{p_0}\right) + p_1 \ln\left(\frac{1}{p_1}\right)\right]. \tag{11.25}$$

To summarize, the relation $\gamma = \lambda(F_{\mathrm{R}}) - h_{\mathrm{KS}}(F_{\mathrm{R}})$ is a consequence of three central results:

1. The existence of a fractal repeller;
2. The existence of a measure on the repeller that can be used for the calculation of dynamical properties; and
3. A positive Lyapunov exponent and KS entropy for trajectories on the repeller, so that when restricted to the repeller, the dynamics of the system is chaotic.

11.5 The escape-rate formula for hyperbolic systems

When this theory is generalized to more complicated systems in higher number of dimensions, the expression for the escape rate becomes, for Anosov systems,

$$\gamma = \sum_{\lambda_i > 0} \lambda_i(F_{\mathrm{R}}) - h_{\mathrm{KS}}(F_{\mathrm{R}}), \tag{11.26}$$

where the sum is over all positive Lyapunov exponents on the repeller.†

While the general derivations of the escape-rate formula, (11.26), are too involved to be presented here in any detail, there is a nice simple argument that captures the general features of the proofs. Let us imagine that we want to determine the rate of information production by trajectories on the repeller. The quantity we want is then $\exp[h_{\mathrm{KS}}(F_{\mathrm{R}})t]$. Using the arguments of the previous chapter, we would expect that this information is generated, exponentially in time, by the stretching of phase-space regions, but reduced, exponentially in time, by the fact that most of the trajectories lead to escape from the system. We would then write

$$e^{h_{\mathrm{KS}}(F_{\mathrm{R}})t} = e^{-\gamma t} \exp\left(t \sum_{\lambda_i > 0} \lambda_i(F_{\mathrm{R}})\right). \tag{11.27}$$

† As we will see in Chapter 13, the Lyapunov exponents and KS entropy appearing in (11.26) should be computed using a Gibbs measure, using the logarithms of the magnitudes of the derivatives of the map in unstable directions to define the appropriate Gibbs measure, as is done in Sec. 11.6, below.

The escape-rate formula follows immediately. More rigorous derivations of this formula are given in papers listed at the end of the chapter.

11.6 Thermodynamic formalism for chaos

The previous results for the escape rate, the Lyapunov exponent, and the KS entropy for the repeller can be combined conveniently into one expression from which all of the others can be derived. This expression is called a *dynamic partition function*. We will not present here the general theory of such partition functions, but instead we shall show how this function can be used to compute the properties of the *one-dimensional* repeller discussed above.

Once again we start with the set of intervals that survive the first n iterations of the map, and we now define a dynamical partition function by

$$Z_n(\beta) = \sum_{i=1}^{2^n} (l_i)^\beta, \tag{11.28}$$

where, as before, the l_i are the lengths of the intervals, and β is to be regarded as a parameter for the moment. We can make this expression look like a canonical partition function if we associate an 'energy', ϵ_i, with each interval l_i ($l_i < 1$), by $l_i = \exp(-\epsilon_i)$. For the simple map that we just discussed, we can compute $Z_n(\beta)$ easily, as

$$\begin{aligned} Z_n(\beta) &= \sum_{n'=0}^{n} \frac{n!}{n'!(n-n')!}(p_0^{n'} p_1^{n-n'})^\beta \\ &= (p_0^\beta + p_1^\beta)^n. \end{aligned}$$

In the same way as one defines a free energy from the equilibrium canonical partition function, we define a quantity called the *topological pressure* from the dynamical partition function. We write $Z_n(\beta)$ as

$$Z_n(\beta) = e^{n\Phi(\beta)}.$$

For large n we can formally compare n with the number of particles N in a canonical ensemble. Then $\Phi(\beta)$ can be thought of as a kind of negative 'free energy per particle times β' and Φ is called the topological pressure, defined properly by the limit

$$\Phi(\beta) = \lim_{n\to\infty} \frac{1}{n} \ln(Z_n(\beta)). \tag{11.29}$$

For our explicit case, we can easily evaluate the topological pressure as

$$\Phi(\beta) = \ln(p_0^\beta + p_1^\beta). \tag{11.30}$$

Now, all of the quantities discussed earlier in this chapter can be expressed in terms of $\Phi(\beta)$. We note that the fractal dimension of the repeller is a solution of $\Phi(d) = 0$, i.e., $p_0^d + p_1^d = 1$. Also $\lambda(F_{\mathrm{R}})$, γ and $h_{\mathrm{KS}}(F_{\mathrm{R}})$ can all be related to $\Phi(\beta)$. Thus, we see that

$$\frac{d\Phi}{d\beta} = \frac{1}{p_0^\beta + p_1^\beta}\left[\ln(p_0)e^{\beta\ln(p_0)} + \ln(p_1)e^{\beta\ln(p_1)}\right],$$

which leads to the result that

$$-\Phi'(\beta)\big|_{\beta=1} = -\frac{p_0}{p_0+p_1}\ln(p_0) - \frac{p_1}{p_0+p_1}\ln(p_1) = \lambda(F_{\mathrm{R}}). \tag{11.31}$$

This holds for more general maps than the one considered above. The escape rate, γ, is

$$\gamma = -\Phi(1) = -\ln(p_0 + p_1). \tag{11.32}$$

The Kolmogorov–Sinai entropy can be written as

$$h_{\mathrm{KS}} = -\Phi'(1) + \Phi(1). \tag{11.33}$$

All of these results are examples of the utility of the dynamical partition function for combining properties of the repeller into one basic expression. The thermodynamic formalism, created by Sinai, Ruelle, and Bowen, provides very deep insights into the structure of chaos and into the connections of dynamical systems theory with statistical mechanics.

11.7 Further reading

We followed here the presentation in the article by Gaspard and Baras [GB92]. See also the discussion in Ott [Ott92], where one can also find a discussion of the various dimensions that characterize fractal sets. The rigorous theory for the dimensions of fractal sets, with interesting examples, is provided in the books by Falconer [Fal93, Fal97] and by Pesin [Pes97].

Derivations of the escape-rate formula can be found in papers by Kadanoff and Tang [KT84], by Tel [Tél90], by Grebogi *et al.* [GOY88], by Kantz and Grassberger [KG85], by Bohr and Rand [BR87], and by Gaspard and Dorfman [GD95].

Detailed and interesting studies of the mathematical properties of repellers have been carried out by Pianigiani and Yorke [PY79], and more recently by Chernov and Markarian [CM97a, CM97b].

The book by Beck and Schlögl [BS93] has a very good introduction to the thermodynamic formalism, and should be read before attempting the more formidable monograph by Ruelle [Rue78].

11.8 Exercises

1. Consider the map described in Exercise 8.1. Suppose $\alpha = \beta = 1/2$. Compute the diffusion coefficient of the map, the escape rate from a lattice of length L, the Lyapunov exponent on the repeller, and the KS entropy on the repeller. You may use the escape-rate formula, and suppose that $L \gg 1$.

2. Consider the Smale horseshoe repeller described in Sec. 11.2. Show that the points on the invariant repeller are isomorphic to Bernoulli sequences of zeros and ones, and that the dynamics on the repeller is isomorphic to the Bernoulli shift. Show also that for the repeller, $\lambda_+ = \ln \eta$ and $h_{\mathrm{KS}} = \ln 2$, by generalizing the construction of Sec. 11.1 to two dimensions, and calculate the box-counting dimension of the repeller.

Use (11.19,11.20) to compute the Hausdorff dimension of the fractal repeller discussed in Sec. 11.1, and show that this result coincides with the box-counting dimension.

12
Transport coefficients and chaos

We have now arrived at a point where we can begin to see what all of the discussions in the previous chapters are leading to. That is, we can now make connections between the dynamical and transport properties of Anosov systems. In this chapter, we discuss two new approaches to the statistical mechanics of irreversible processes in fluids that use almost all of the ideas that we have discussed so far. These are the escape-rate formalism of Gaspard and Nicolis, and the Gaussian thermostat method due to Nosé, Hoover, Evans and Morriss. It should be mentioned at the outset that this is a new area of research, that many more developments can be expected from this approach to transport, and that what we will discuss here are merely the first glimmerings of the results that can be obtained by thinking of transport phenomena in terms of the chaotic properties of reversible dynamical systems. There is a third, closely related, dynamical approach to transport coefficients based upon the properties of unstable periodic orbits of a hyperbolic system. We will discuss this approach in Chapter 15.

12.1 The escape-rate formalism

Suppose we think of a system that consists of a particle of mass m and energy E, moving among a fixed set of scatterers which are in some region R which is of infinite extent in all directions except one, the x-direction, such that the scatterers are confined to the interval $0 \leq x \leq L$. Absorbing walls are placed at the (hyper)planes at $x = 0$ and $x = L$ (see Fig. 12.1). For large L and for large times after some initial time, we expect the spatial distribution function, $P(\mathbf{r}, t)$ of the moving particle to be described by the diffusion equation, $\partial P/\partial t = D\nabla^2 P$, where D is the dif-

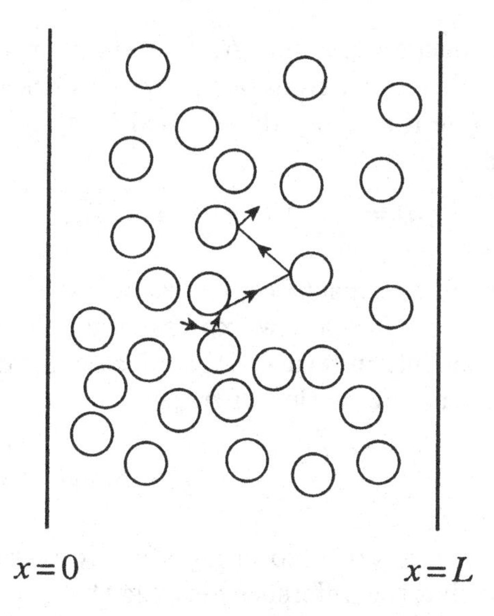

Fig. 12.1 The slab geometry for diffusion in a Lorentz system with absorbing boundaries.

fusion coefficient. The absorbing boundary conditions lead to the condition that $P(r,t)|_{\text{boundary}} = 0$. Then the probability for the distribution of particles in the x-direction has the form:

$$P(x,t) = \sum_{n=1}^{\infty} a_n \sin\left(\frac{\pi n}{L}x\right) \exp\left(-\frac{(n\pi)^2}{L^2}Dt\right),$$

where the a_n are numerical coefficients. The decay of the probability distribution for the moving particle is exponential; the slowest decaying mode, $n = 1$ decays as $\exp(-\pi^2 Dt/L^2)$. So for large systems we can define a *macroscopic escape rate* as

$$\gamma_{\text{mac}} = \frac{\pi^2}{L^2}D. \tag{12.1}$$

Now let's consider this process from a microscopic point of view using the escape-rate method discussed earlier. If we think of the set of trajectories of the moving particle inside the region R, starting with all possible initial positions and velocities, then all but a set of measure zero of the initial phases of the moving particle will lead to absorption of the particle by the walls. There will remain a fractal set of initial phases for which the particle will be trapped forever inside the region R. This set of trajectories forms the repeller F_{R}. We know from our previous work that there is a microscopic escape rate, γ_{mic}, determined by the fractal repeller, given by (11.26),

$$\gamma_{\mathrm{mic}} = \sum_{\lambda_i > 0} \lambda_i(F_{\mathrm{R}}) - h_{\mathrm{KS}}(F_{\mathrm{R}}). \tag{12.2}$$

If we now identify the macroscopic escape rate with the microscopic escape rate we find a new expression for the diffusion coefficient D given entirely in terms of the microscopic quantities that characterize the chaotic motion of trajectories on the repeller

$$D = \lim_{L\to\infty} \frac{L^2}{\pi^2} \left(\sum_{\lambda_i > 0} \lambda_i(F_{\mathrm{R}}) - h_{\mathrm{KS}}(F_{\mathrm{R}}) \right). \tag{12.3}$$

The limit $L \to \infty$ is taken to eliminate any possible finite-size effects. We note that the difference between the sum of the positive Lyapunov exponents and the KS entropy must be of order L^{-2} in order that the limit in (12.3) exist.

This formula has been applied so far to two-dimensional Lorentz gases, where the scatterers are hard disks placed on a regular lattice in a plane, as well as for Lorentz gases in two and three dimensions where the scatterers are disks or spheres placed at random in the plane or in space. It has also been applied to determine the diffusion coefficient for particles whose dynamics are governed by discrete maps in one or two dimensions. This new, deterministic way of describing transport phenomena can also be applied to obtain analogous results for the coefficients of thermal conductivity and viscosity as well as other transport and chemical reaction rate coefficients.

12.2 Gaussian thermostats

In attempting to simulate transport processes in fluids on a computer, Nosé, Hoover, Evans, Morriss and others noticed that imposed gradients in velocity or temperature or external gravitational or electric fields led to noticeable changes in the energy of the system. For example, the shearing needed to generate a viscous flow leads to considerable viscous heating of the fluid. This heating tended to interfere with the effects that these workers wished to study. To deal with this heating problem, they used an internal 'thermostat' – a fictitious frictional force – which maintained the energy (total, or the kinetic energy) at some constant pre-determined value. They thereupon had to develop a statistical mechanics of systems subjected to such thermostats. Among other results of this new statistical mechanics is a connection between transport coefficients and Lyapunov exponents of the thermostatted systems that is also of considerable interest and the object of much current study.

We give an application of this method as it applies to the calculation of the electrical conductivity for a charged particle moving in a two-dimensional array of fixed hard disk scatterers. An electric field is applied to the system which only affects the moving particle.

At an elastic collision with a scatterer, the momentum of a particle changes direction according to (2.6),

$$\begin{aligned} \mathbf{p}' &= \mathbf{p} - 2\left(\mathbf{p}\cdot\hat{\mathbf{k}}\right)\hat{\mathbf{k}}, \\ \mathbf{p}^2 &= \mathbf{p}'^2. \end{aligned}$$

In between collisions, an electric field, $\mathbf{E}$, affects the trajectories of particles and accelerates them in the direction of the field. The usual equations of motion $\dot{\mathbf{r}} = \mathbf{p}$, $\dot{\mathbf{p}} = q\mathbf{E}$ $(m = 1)$ lead to an increase in the average kinetic energy of the moving particle. A fictitious thermostat is then applied to the moving particle which is designed to keep its kinetic energy constant. Thus, Newton's equations of motion must be modified to include the effects of the thermostat. We suppose that the equation for the position of the particle, $\dot{\mathbf{r}} = \mathbf{p}$, remains unaltered, but we add a friction term to the momentum equation, $\dot{\mathbf{p}} = q\mathbf{E} - \alpha(\mathbf{p})\mathbf{p}$, where the friction coefficient $\alpha(\mathbf{p})$ is fixed by the condition that the kinetic energy stays constant:

$$\frac{d\mathbf{p}^2}{dt} = 0.$$

This condition is equivalent to $\mathbf{p}\cdot\dot{\mathbf{p}} = 0$, so that

$$\mathbf{p}\cdot\dot{\mathbf{p}} = \mathbf{p}\cdot q\mathbf{E} - \alpha p^2 = 0,$$

that is,

$$\alpha(\mathbf{p}) = \frac{q\mathbf{E}\cdot\mathbf{p}}{p^2}.$$

It is interesting to note that Gauss formulated a principle of least constraint for systems such as this which leads to the same equation.

The equations of motion are now

$$\begin{aligned} \dot{\mathbf{r}} &= \mathbf{p}, \\ \dot{\mathbf{p}} &= q\mathbf{E} - \frac{q\mathbf{E}\cdot\mathbf{p}}{p^2}\mathbf{p}. \end{aligned} \tag{12.4}$$

The utility of this approach is that the frictional force can be simulated on a computer. Note that the equations of motion are still time-reversible ($\mathbf{p} \to -\mathbf{p}$, $t \to -t$), because $\alpha(\mathbf{p}) = -\alpha(-\mathbf{p})$. Because of the friction term, Liouville's equation has to be changed, too. Generally,

$$\frac{\partial \rho}{\partial t} + \frac{\partial}{\partial \mathbf{r}} \cdot (\dot{\mathbf{r}}\rho) + \frac{\partial}{\partial \mathbf{p}} \cdot (\dot{\mathbf{p}}\rho) = 0$$

expresses the conservation of systems in Γ-space. Using the chain rule, we have

$$\frac{\partial \rho}{\partial t} + \dot{\mathbf{r}} \cdot \frac{\partial}{\partial \mathbf{r}}\rho + \dot{\mathbf{p}} \cdot \frac{\partial}{\partial \mathbf{p}}\rho + \rho\left[\frac{\partial \dot{\mathbf{r}}}{\partial \mathbf{r}} + \frac{\partial \dot{\mathbf{p}}}{\partial \mathbf{p}}\right] = 0.$$

The terms in the brackets normally cancel, but not in this case because of the friction; instead

$$\frac{\partial \dot{\mathbf{r}}}{\partial \mathbf{r}} + \frac{\partial \dot{\mathbf{p}}}{\partial \mathbf{p}} = -\frac{\partial}{\partial \mathbf{p}} \cdot (\alpha \mathbf{p}). \tag{12.5}$$

This means, among other things, that phase-space volumes are *not* preserved in time, but change according to the dynamical evolution of the system.† We can easily compute the right-hand side of (12.5):

$$\begin{aligned} \frac{\partial}{\partial \mathbf{p}} \cdot (\alpha \mathbf{p}) &= \frac{\partial}{\partial p_x}\left(\frac{q\mathbf{E}\cdot\mathbf{p}}{p^2}p_x\right) + \frac{\partial}{\partial p_y}\left(\frac{q\mathbf{E}\cdot\mathbf{p}}{p^2}p_y\right) \\ &= \frac{2q\mathbf{E}\cdot\mathbf{p}}{p^2} + \frac{qE_x p_x}{p^2} - \frac{2qp_x(\mathbf{E}\cdot\mathbf{p})p_x}{(p_x^2+p_y^2)^2} \end{aligned}$$

† The reader may note that we have not explicitly included the effects of the instantaneous, specular collisions of the moving particle with the scatterers in these equations . It is not difficult to include these effects and to show that the final result obtained below, (12.11), is unchanged. We will not elaborate on this point here.

$$+\frac{qE_yp_y}{p^2}-\frac{2qp_y(\mathbf{E}\cdot\mathbf{p})p_y}{(p_x^2+p_y^2)^2}$$
$$=\frac{3q(\mathbf{E}\cdot\mathbf{p})}{p^2}-\frac{2q\mathbf{E}\cdot\mathbf{p}}{p^2}$$
$$=\alpha. \tag{12.6}$$

The modified Liouville equation is now

$$\frac{\partial\rho}{\partial t}+\dot{\mathbf{r}}\cdot\frac{\partial}{\partial\mathbf{r}}\rho+\dot{\mathbf{p}}\cdot\frac{\partial}{\partial\mathbf{p}}\rho-\rho\alpha=0, \tag{12.7}$$

or

$$\frac{d\rho}{dt}=\alpha\rho. \tag{12.8}$$

We are now going to give an argument that relates the entropy production in this thermostatted system to the average value of α. First of all, this argument has some serious flaws when applied to a nonequilibrium steady state, for a reason that will be discussed after we have obtained a conclusion. It turns out that the conclusion we will reach is indeed correct, even though the intermediate argument is problematic. To do this calculation properly for a nonequilibrium steady state, one has to use the notion of SRB measures, which are discussed in the next chapter.

We begin by defining the entropy of the moving particle in the thermostat (or rather, of an ensemble of moving particles) by applying the definition of entropy suggested by the H-theorem to a system with a Gaussian thermostat:

$$S_{\mathrm{G}}\equiv-k_{\mathrm{B}}\int d\mathbf{r}\int d\mathbf{p}\rho(\mathbf{r},\mathbf{p},t)[\ln\rho(\mathbf{r},\mathbf{p},t)-1]. \tag{12.9}$$

Then

$$\frac{dS_{\mathrm{G}}}{dt}=-k_{\mathrm{B}}\int d\Gamma[\ln\rho]\frac{\partial\rho}{\partial t}$$
$$=k_{\mathrm{B}}\int d\Gamma[\ln\rho]\left[\frac{\partial}{\partial\mathbf{r}}\cdot(\dot{\mathbf{r}}\rho)+\frac{\partial}{\partial\mathbf{p}}\cdot(\dot{\mathbf{p}}\rho)\right]. \tag{12.10}$$

After an integration by parts, we find

$$\frac{dS_{\mathrm{G}}}{dt}=-k_{\mathrm{B}}\int d\Gamma\left[\dot{\mathbf{r}}\cdot\frac{\partial\rho}{\partial\mathbf{r}}+\dot{\mathbf{p}}\cdot\frac{\partial\rho}{\partial\mathbf{p}}\right],$$

and another partial integration yields

$$\frac{dS_{\mathrm{G}}}{dt}=k_{\mathrm{B}}\int d\Gamma\rho\left(\frac{\partial\dot{\mathbf{r}}}{\partial\mathbf{r}}+\frac{\partial\dot{\mathbf{p}}}{\partial\mathbf{p}}\right)$$

$$= -k_{\mathrm{B}} \int d\Gamma \rho \alpha = -k_{\mathrm{B}} \langle \alpha \rangle . \qquad (12.11)$$

We expect that the average friction coefficient is positive, so we find that the entropy is a *decreasing* function of time. We can argue that we need to consider both the system and the thermostat when computing the total entropy change of the universe. Since the thermostat acts as a reservoir, the entropy of the reservoir has to increase to compensate for the decrease of the system's entropy:

$$\frac{dS_{\text{reservoir}}}{dt} \geq k_{\mathrm{B}} \langle \alpha \rangle \geq 0. \qquad (12.12)$$

The entropy production in the reservoir is then taken to represent the macroscopic entropy production, and is related to the electrical conductivity of our system, by the equations of irreversible thermodynamics:

$$\frac{dS_{\text{reservoir}}}{dt} = \frac{\mathbf{J} \cdot \mathbf{E}}{T} = \frac{\sigma \mathbf{E}^2}{T}, \qquad (12.13)$$

where $\mathbf{J}$ is the average electrical current, $\mathbf{J} = \sigma \mathbf{E}$, and σ is the electrical conductivity. In a computer simulation, the system reaches a steady state, where there is electrical conduction without a change in energy, and we suppose that the equality in (12.12) holds:

$$k_{\mathrm{B}} \langle \alpha \rangle = \frac{\sigma E^2}{T}. \qquad (12.14)$$

Finally, we arrive at an expression for the electrical conductivity,

$$\sigma = \frac{k_{\mathrm{B}} T \langle \alpha \rangle}{E^2}. \qquad (12.15)$$

This relates the average friction coefficient that can be measured in computer simulations to the electrical conductivity of the system. The flaw in this argument is that we always assume that the phase-space density ρ is a nice smooth function with smooth derivatives. What is really happening as the system approaches a steady state from some initial condition is that the phase-space density is becoming concentrated on an attractor. In the steady state, the density on the attractor is not an smooth function anymore and the integral on the right-hand side of (12.10) has no meaning. Nevertheless, one can indeed do a more careful analysis of the entropy production in terms of the proper mathematical objects, SRB measures, and still recover our main result which is (12.18), below. We will make some additional remarks on this point at the end of this section.

The connection to chaotic dynamics is established by using our earlier result that phase-space volumes are not conserved by the thermostatted dynamics. The change in phase-space volumes can then be related to a change in Lyapunov exponents which no longer satisfy the symplectic condition

$$\lambda_+ + \lambda_- = 0 \tag{12.16}$$

for our system which has only two non-zero Lyapunov exponents.†

We now relate the expression for electrical conductivity, (12.15), to the Lyapunov exponents for the thermostatted systems. To determine the rate of change of a phase-space region, imagine that we have a fixed number $\mathcal{N}$ of members of the ensemble in some small volume of phase-space $\mathcal{V}(t)$. Then the phase-space density is $\rho = \mathcal{N}/\mathcal{V}(t)$. If we now use the modified Liouville equation, (12.7), we find that

$$-\frac{\mathcal{N}}{\mathcal{V}^2}\frac{d\mathcal{V}}{dt} = \frac{\alpha\mathcal{N}}{\mathcal{V}},$$

or,

$$\frac{d\mathcal{V}(t)}{dt} = -\alpha\mathcal{V}.$$

If, as we expect, $\alpha > 0$, the volume in phase-space will decrease, this time not because of escape, but because of the effects of the thermostat. If we now consider the steady-state ensemble average of this equation we find

$$\left\langle \frac{d\ln\mathcal{V}}{dt} \right\rangle = -\langle\alpha\rangle. \tag{12.17}$$

Now the change in the phase-space volume is associated with the sum of the positive and negative Lyapunov exponents for the system, so that (12.17) can be written

$$\langle\lambda_+ + \lambda_-\rangle = -\langle\alpha\rangle.$$

† For the case of a point particle moving among fixed disks in a plane, the phase-space of the system is four-dimensional. The requirement of constant kinetic energy reduces the phase-space to a three-dimensional constant-energy surface for this model. Then there is one zero Lyapunov exponent connected with the direction along a phase-space trajectory. This leaves at most two non-zero exponents.

This allows us to express the electrical conductivity in terms of the average Lyapunov exponents for the system, in the steady state, as

$$\sigma = -\frac{k_B T}{E^2} \langle \lambda_+ + \lambda_- \rangle . \tag{12.18}$$

If the system is ergodic, the average values on the right-hand side of (12.18) may be removed and a clear relationship between the electrical conductivity and the Lyapunov exponents follows. Similar results for other transport coefficients have been obtained.

We conclude this discussion with a few remarks.

1. The thermostat produces a contraction of the phase-space onto an attractor of lower dimension than the phase-space itself. In fact, this contraction is somewhat subtle. Quite often the box-counting dimension of the attractor is the same as the dimension of the phase-space itself. However, there are other measures of dimension of a set, in addition to the box-counting dimension, such as the Hausdorff dimension discussed in Chapter 11, which are more sensitive to the fractal nature of the attractor. It is by means of these other measures of the dimension that one can see the appearance of the fractal attractor in the steady state. Typically, one considers the information dimension which does indeed characterize the contraction of the system onto an attractor. We will discuss an example of this in the next chapter.

The fractal structures that appear for this kind of an attractor are quite different from that for the repeller underlying transport. As we will see in the next chapter where we study SRB and Gibbs measures, the typical attractor of a thermostatted system has a measure which is smooth in the unstable directions and fractal along the stable directions. For a repeller, the measure has fractal properties in both the stable and unstable directions, and is not an SRB measure. Instead one needs a more general measure, called a Gibbs measure, to describe it.

2. Gaspard, for open Hamiltonian systems, and Tél *et al.*, for thermostatted systems, have used the idea of a coarse-grained entropy to argue that on a coarse-grained scale, there is an *increase* of entropy. The fundamental idea is that the dynamics on an attractor or repeller takes place on an infinitesimally fine scale, as will be discussed in the next chapter. On a coarse-grained scale, this dy-

namics is unobservable, and information about the system is lost. This loss of information appears to the observer as a positive rate of entropy production. The magnitude of the entropy depends on the resolution of the graining, but the rate of entropy production does not. In the case of thermostatted systems, Gilbert and Dorfman have shown that the coarse-graining method leads to (12.18) above. Goldstein, Lebowitz, and Sinai have reached similar conclusions by a different and elegant method. We refer the reader to the papers listed below for further details.
3. In order for the electrical conductivity to be well-behaved for small electric fields, the sum of the two Lyapunov exponents must be of order E^2 for small E. This too is reminiscent of the L^2 dependence of the transport coefficients in the escape-rate formula, (12.3). While one recovers the same values for the transport coefficients in the limit as $E \to 0$ as one obtains by evaluating the Green–Kubo formulae, it remains to be determined what physical meaning one can ascribe to the field-dependent transport coefficients in thermostatted systems for general values of the external field.
4. Recent work using methods to be described in Chapter 15 has led to theoretical values for the Lyapunov exponents for a random Lorentz gas in a thermostatted electric field. These results are totally consistent with the picture developed here and allow one to make this formalism more concrete for simple systems, at least.

12.3 Further reading

The escape-rate formalism for transport coefficients was initiated by Gaspard and Nicolis [GN90], and some further papers are by Gaspard and Baras [GB92, GB95], Dorfman and Gaspard [DG95], and Gaspard and Dorfman [GD95].

The method of studying transport in thermostatted systems was developed by Evans and co-workers [EM90], and by Hoover and co-workers [Hoo91]. Papers that develop the theory further are by Chernov *et al.* [CELS93], Evans *et al.* [ECM90], Gallavotti and Cohen [GC95b], Tél *et al.* [TVB96] and by Morriss and Rondoni [MR96]. The most up-to-date collections of papers on this subject are collected in an issue of *Physica A* edited by Maréschal [Mar97] and in an issue of *Chaos* edited by Tél, Gaspard, and

Nicolis [TGN98]. The original Gauss principle of least constraint is described in the classical text by Whittaker [Whi88] in Section 105.

Discussions of entropy-related issues can be found in the above-listed papers as well as in presentations of Gaspard [Gas97b, Gas98], papers by Tél *et al.* [TVB96, BTV96, VTB97, BTV98], by Ruelle [Rue96a], by Gallavotti and Ruelle [GR97], by Gilbert and Dorfman [GD98], and by Goldstein *et al.* [GLS98].

Discussions of the dimensions of fractal sets are quite common in the books on dynamical systems. For more on this matter see the books by Pesin [Pes97], by Falconer [Fal97], by Ott [Ott92], by Alligood *et al.* [ASY97], and by Beck and Schlögl [BS93].

13
Sinai–Ruelle–Bowen (SRB) and Gibbs measures

We have just seen that in a system subjected to an external field and a thermostat which maintains the system's kinetic energy at a constant value there is a phase space contraction taking place. Prior to that, we showed that a fractal repeller can form in the phase-space for a Hamiltonian system with absorbing boundaries such that trajectories hitting the boundary never re-appear in the system. Some questions naturally arise as we think about these systems, such as:

1. To what kind of structure is the phase-space for a thermostatted system contracting?
2. What are the properties of such a structure? Is it a smooth hypersurface, say, or does it have a more complicated structure?
3. How do we describe fractal attractors and/or repellers and compute the properties of trajectories which are confined to them, since these objects are typically of zero Lebesgue measure in phase-space?

In this chapter, we will show that for each of these situations there is an appropriate measure that can be used to describe the resulting sets and to compute the properties of trajectories which are confined to these sets. Throughout our discussions we will suppose that the dynamics is hyperbolic. In the thermostatted case, the system contracts onto an attractor, and the attractor is characterized by an invariant measure which is smooth along unstable directions and fractal in the stable directions. Such measures are known as *SRB (Sinai–Ruelle–Bowen) measures.*

For a system with escape, the invariant measure on the repeller is different from that on an attractor, because escape takes place along the expanding directions. Consequently, the invariant measure will not be smooth in the expanding direction, as we saw already in the case of the one-dimensional repellers, studied in Chapter 11. However, there is a measure on the repeller, called a *Gibbs measure*, that adequately describes the invariant state. Gibbs measures may also be constructed in a more general context than systems with repellers, and from this point of view, an SRB measure is a special case of a Gibbs measure with smoothness properties in the expanding directions. Here we describe, in an elementary way, the properties of such measures. We begin with SRB measures.

13.1 SRB measures

In the previous chapter, we discussed the motion of a charged particle moving in an array of fixed scatterers in an electric field with an imposed thermostat which keeps the kinetic energy of the moving particle constant. There we found that there is an average contraction of phase-space volumes, that is, the sum of the Lyapunov exponents is negative, rather than zero, as would be the case for a Hamiltonian system. Moreover, the equations of motion for the moving particle are reversible, that is, they are invariant under the transformation $t \to -t$, $\mathbf{p} \to -\mathbf{p}$, as they would be also for a Hamiltonian system. How can it be, then, that this system has contracting phase-space volumes, which seems to imply a direction in time, while the equations of motion are invariant under the usual time reversal operation?

The answer to this question is somewhat subtle but illuminating, and to discover it we look, as usual, at a very simple model system for insight into this situation. We consider a slight generalization of the baker's transformation on the unit square (see Fig. 13.1). Instead of expanding and contracting areas equally, we expand one region and contract the other while still preserving the overall area of the square. Let us consider then, the following

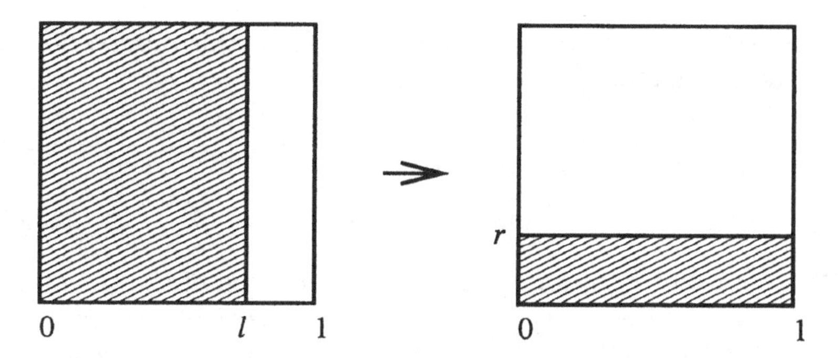

Fig. 13.1. Illustration of the map in Eq. (13.1).

transformation, M,

$$\begin{pmatrix} x' \\ y' \end{pmatrix} = M \cdot \begin{pmatrix} x \\ y \end{pmatrix} = \begin{cases} \begin{pmatrix} x/l \\ ry \end{pmatrix} & \text{for } 0 \leq x \leq l, \\ \begin{pmatrix} (x-l)/r \\ r+ly \end{pmatrix} & \text{for } l < x \leq 1. \end{cases} \tag{13.1}$$

Here we set $r + l = 1$, but suppose that $r \neq l$. Thus one side of the unit square has its area expanded while the area of the other is contracted in such a way that the total area remains constant. The Jacobian of the transformation is not unity but instead takes on two values depending upon whether $x < l$ or $x > l$. That is,

$$\left| \frac{\partial(x', y')}{\partial(x, y)} \right| = \begin{cases} r/l & \text{for } 0 \leq x \leq l, \\ l/r & \text{for } l < x \leq 1. \end{cases} \tag{13.2}$$

The positive and negative Lyapunov exponents for this map may easily be determined by a simple argument. Consider two infinitesimally close points somewhere in the unit square. Then the expanding direction is in the x-direction, and the contracting direction is the y-direction. After n steps, the separation of the points in the unstable direction will be

$$\delta x(n) = l^{-n_1} r^{-n_2} \delta x(0), \tag{13.3}$$

and their separation in the stable direction will be

$$\delta y(n) = r^{n_1} l^{n_2} \delta y(0). \tag{13.4}$$

Here, n_1 is the number of times the nearby points find themselves in the region $0 \leq x \leq l$, and n_2 is the number of times they are in the region $l < x \leq 1$, with $n_1 + n_2 = n$. From this observation it immediately follows that

$$\lambda_+ = \lim_{n\to\infty} \left[\frac{n_1}{n} \ln \frac{1}{l} + \frac{n_2}{n} \ln \frac{1}{r} \right], \tag{13.5}$$

$$\lambda_- = \lim_{n\to\infty} \left[\frac{n_1}{n} \ln r + \frac{n_2}{n} \ln l \right]. \tag{13.6}$$

Here, the ratios n_i/n for $i = 1, 2$ are the fraction of the time that the points find themselves to the left or to the right, respectively, of the line $x = l$. If the map were ergodic, these ratios would be equal to the measures of the regions on each side of this line. One might guess, and as we show more carefully below, the measures of the regions to the left and to the right of the line $x = l$ are equal to l and to $1 - l = r$, respectively, so that the Lyapunov exponents are

$$\begin{aligned} \lambda_+ &= l \ln l^{-1} + r \ln r^{-1} \\ \lambda_- &= l \ln r + r \ln l, \end{aligned} \tag{13.7}$$

and the sum of the two Lyapunov exponents is negative, except for the case when $l = r = 1/2$, where the sum is zero,

$$\lambda_+ + \lambda_- = (l - r) \ln(r/l) \leq 0. \tag{13.8}$$

The case where the sum is zero is, of course, the baker's transformation. We note that for any other value of l, the map contracts the phase-space area, since the sum of the Lyapunov exponents is negative. However, we have as yet no clear idea what this means since the unit square is always reconstructed at each iteration of the map. The resolution of this question will depend upon the observation that a set may have many different dimensions, in addition to the box-counting dimension discussed earlier. We will see that the set which is being constructed by this map is one such set with many dimensions, a so-called *multifractal* set.

Before going into the nature of the phase-space contraction, we note one more property of the map. It is, in fact, reversible! That is, there is a transformation Φ for this map, which plays the same role as the time-reversal transformation, T, for mechanical systems. To understand this, we need to look at what happens in a mechanical system under time-reversal. In our discussion of linear response theory in Chapter 6, we constructed a time evolution operator $\exp(t\mathcal{L})$ which acts on dynamical functions to give their values at a time t later, that is,

$$e^{t\mathcal{L}} A(\Gamma) = A(\Gamma_t), \tag{13.9}$$

where Γ is some phase point, and Γ_t is the point to which Γ evolves after a time t. A time-reversal operator, T, is defined in mechanics as one that has the properties that

$$\begin{aligned} Te^{t\mathcal{L}}T &= \left(e^{t\mathcal{L}}\right)^{-1} = e^{-t\mathcal{L}}, \qquad (13.10) \\ TT &= T^2 = 1. \qquad (13.11) \end{aligned}$$

For a simple mechanical system consisting of a collection of particles interacting with velocity-independent potentials, one can easily show that the operator that replaces $\mathbf{p}_i$ by $-\mathbf{p}_i$ for every momentum variable appearing to the right of it is the time-reversal operator T. Time-reversal invariance in classical mechanics is often expressed by the equivalent statement to those above, namely,

$$e^{t\mathcal{L}}T = Te^{-t\mathcal{L}}. \tag{13.12}$$

Time-reversal invariance for a map M means that the map is invertible, and that there exists an operator Φ that has the properties

$$\begin{aligned} \Phi M \Phi &= M^{-1} \qquad (13.13) \\ \Phi\Phi &= 1, \qquad (13.14) \end{aligned}$$

similar to those for T above. Here M^{-1} is the inverse of the map M. The map M given by (13.1) is invertible with inverse

$$M^{-1} \cdot \begin{pmatrix} x \\ y \end{pmatrix} = \begin{cases} \begin{pmatrix} lx \\ y/r \end{pmatrix} & \text{for } 0 \leq y \leq r, \\ \begin{pmatrix} rx + l \\ (y-r)/l \end{pmatrix} & \text{for } r < y \leq 1. \end{cases} \tag{13.15}$$

A time-reversal operator Φ for the map M is then easily shown to be

$$\Phi \cdot \begin{pmatrix} x \\ y \end{pmatrix} = \begin{pmatrix} 1-y \\ 1-x \end{pmatrix}. \tag{13.16}$$

This time-reversal operator does not depend on l, r and can be used for the baker map as well.

We now have a map which is time-reversible, and with Lyapunov exponents summing to a negative quantity, and thus similar to the situation that one obtains in the dynamics of thermostatted systems discussed earlier. It is now time to construct the SRB measure. We want to find an invariant density (under M) on the unit square that will characterize the final state of some initial distribution evolving under the map M. To do this, we consider the Frobenius–Perron equation, (10.1), for the evolution of the phase-space density function $\rho_n(x, y)$ under map M, where n is the number of time steps after some initial time. For our case, the Frobenius–Perron equation is

$$\begin{aligned} \rho_n(x,y) &= \int_0^1 dx' \int_0^1 dy' \rho_{n-1}(x',y') \times \\ &\qquad \delta(x - M_x(x',y'))\delta(y - M_y(x',y')) \\ &= \begin{cases} (l/r)\rho_{n-1}(lx, y/r) & \text{for } 0 \leq y \leq r, \\ (r/l)\rho_{n-1}(l + rx, (y-r)/l) & \text{for } r < y \leq 1. \end{cases} \end{aligned} \tag{13.17}$$

Here, M_x, M_y denote the x, y components, respectively, of the map M. The stationary state distribution, if it exists, is found by dropping the time dependence of the distribution functions appearing in (13.18)below, and looking for solutions where the same functions appear on the right- and left-hand sides. As noted above, we expect the distribution function to describe a multifractal structure with a variety of dimensions, so we don't expect to obtain a simple analytic solution of the stationary Frobenius–Perron equation. We therefore must resort to some subtle methods to solve it. This is accomplished by means of cumulative distribution functions.

13.2 The cumulative functions

A useful method to treat functions which we suspect might be singular in some way is to consider their integral over some region. The integration may 'smooth out' the singular properties and provide a more analytically tractable function. We try this method here. Consider, then, a cumulative distribution function $G(x, y)$ defined by

$$G(x,y) = \int_0^x \int_0^y dx' dy' \rho(x',y'), \tag{13.18}$$

where $\rho(x,y)$ is a stationary solution of (13.18). An equation for $G(x,y)$ is readily obtained from (13.18) as

$$G(x,y) = \begin{cases} G(lx, y/r) & \text{for } 0 \le y \le r, \\ G(lx,1) + G(rx+l, (y-r)/l) & \\ \qquad - G(l, (y-r)/l) & \text{for } r < y \le 1. \end{cases} \tag{13.19}$$

Note that we can consistently set $G(1,1) = 1$ as a good normalization of the invariant density. Next we see if we can solve this equation by separating the x and y variables. This would correspond to the stationary density function being a product of distributions in the x- and y- directions, separately. Moreover, since the x-direction is the expanding direction, we might expect the distribution in this direction to be smooth, while the analytic problems would be connected with the stable direction, the y-direction, as it is in the baker's map, of which this is a variant. Consequently, we try a solution of the form $G(x,y) = xH(y)$, and find that $H(y)$ must satisfy $H(0) = 0, H(1) = 1$, and

$$H(y) = \begin{cases} lH(y/r) & \text{for } 0 \le y \le r, \\ l + rH((y-r)/l) & \text{for } r < y \le 1. \end{cases} \tag{13.20}$$

This is a lovely equation with a striking solution. To see how a solution might be generated, we consider a recursive solution, where we solve the system of equations

$$H^{(k+1)}(y) = \begin{cases} lH^{(k)}(y/r) & \text{for } 0 \le y \le r, \\ l + rH^{(k)}((y-r)/l) & \text{for } r < y \le 1, \end{cases} \tag{13.21}$$

by starting with a function $H^{(0)}(y)$ that satisfies the boundary conditions, and then generate the $H^{(k)}(y)$ recursively. A suitable function $H^{(0)}(y)$ is $H^{(0)}(y) = 0$ for $0 \le y \le r$ and $H^{(0)}(y) = 1$ for $r < y \le 1$. The result for $H^{(1)}(y)$ follows immediately as

$$H^{(1)}(y) = \begin{cases} 0 & \text{for } 0 \le y \le r^2, \\ l & \text{for } r^2 < y \le r + rl, \\ 1 & \text{for } r + rl < y \le 1. \end{cases} \tag{13.22}$$

Some of the successive approximations are illustrated in Figs. 13.2 to 13.5.

The curious feature of these successive solutions is that they are functions with derivative zero almost everywhere. As one takes the limit $k \to \infty$, one obtains a function which is non-decreasing, yet

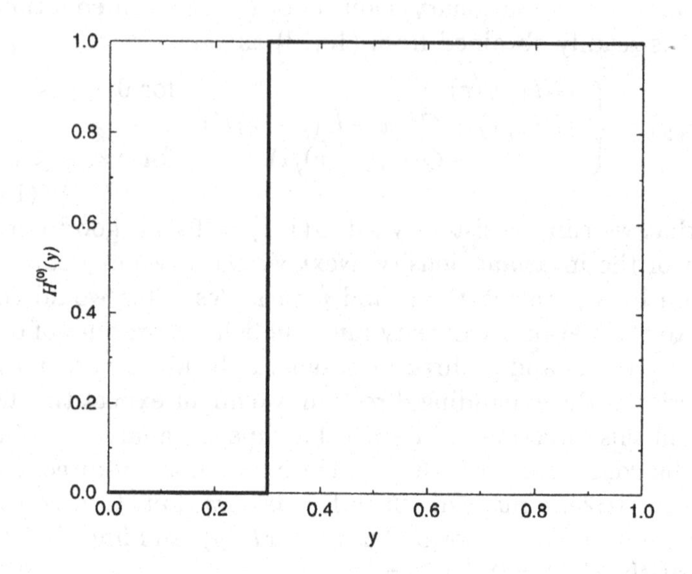

Fig. 13.2 The first approximation, $H^{(0)}(y)$, to the Lebesgue singular function. Here, $r = 0.3$.

with zero derivative almost everywhere, with respect to the ordinary Lebesgue measure. This function is known in the mathematical literature as a *Lebesgue singular function.* Equation 13.19 is an equation of the de Rham type, and it is known that the solution we have constructed is the unique solution of this equation.

The ingredients of the SRB measure for our map are all now in place. Since we know the cumulative function is the product of x and $H(y)$, we can express the ensemble average of any function, $f(x, y)$, on the unit square as an integral over an SRB measure. This average must be expressed as a Lebesgue–Stieltjes integral

$$\int f(x,y)d\mu_{\mathrm{SRB}} = \int_0^1 dx \int_0^1 dH(y)f(x,y). \tag{13.23}$$

We have an invariant SRB measure on the unit square which is smooth along the unstable direction and singular along the stable direction. The initial unit square has contracted to a set with this kind of 'smooth-singular' invariant measure.

We can now look at some corollaries of this construction. First,

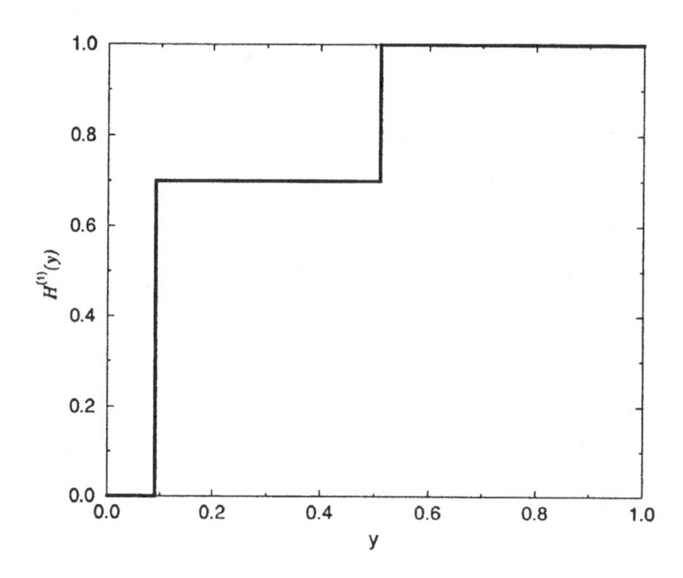

Fig. 13.3 The next approximation, $H^{(1)}(y)$, to the Lebesgue singular function. Here, $r = 0.3$.

we return to the computation of the Lyapunov exponents. For these we needed the invariant measures of the regions $0 \leq x \leq l$ and $l < x \leq 1$. It follows from the fact that the invariant measure is smooth in the x-direction, and integrates to unity in the y-direction, that the invariant measures of these two regions are l and $1 - l = r$, respectively. Although we have not proved that the dynamics of our map is ergodic, a proof can be constructed along the lines of the proof for the baker's map. From this, it follows that our intuitive calculation of the Lyapunov exponents given by (13.5) and (13.6), gives the correct result.

Second, we consider the fractal dimensions of the set whose invariant measure is given by the SRB measure just computed. We want to characterize in some way the dimension of the 'smallest' set in the unit square whose SRB measure is unity. That is, we want the infimum, over S, of the Hausdorff dimensions of all sets S, such that each set S has the full SRB measure of the attractor. The 'smallest' set S is the attractor to which the full unit square

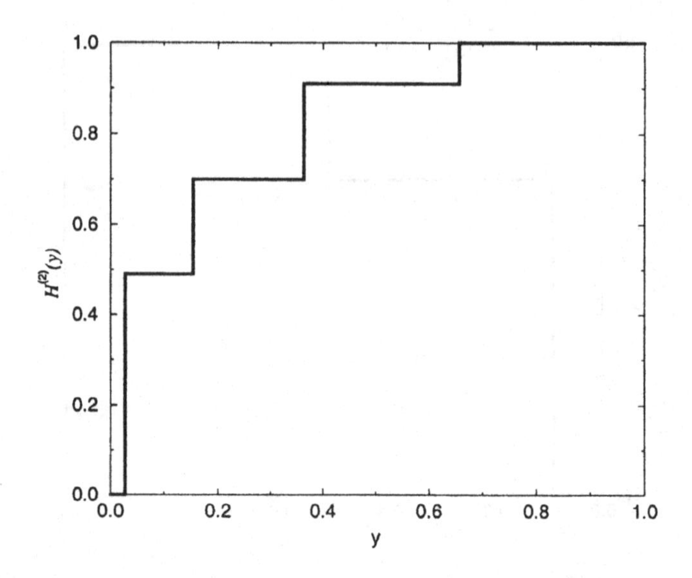

Fig. 13.4. The third approximation, $H^{(2)}(y)$. Here, $r = 0.3$.

contracts under the action of the map M. As such, it is the analog of the attractor that appears in the phase-space of a thermostatted system. To get some intuitive feeling for this set, let's consider the support on the unit square of the approximate measures generated by the cumulative functions $G^{(k)}(x, y) = xH^{(k)}(y)$. Since the phase-space density associated with $G^{(k)}(x, y)$ is simply its derivatives with respect to x and y,

$$\rho^{(k)}(x, y) = \frac{\partial^2}{\partial x \partial y} G^{(k)}(x, y), \tag{13.24}$$

we see that $\rho^{(k)}(x, y)$ is uniform in the x-direction and is a sum of delta functions with various weights, at the points of discontinuity of $H^{(k)}(y)$. That is to say, the support of $\rho^{(k)}(x, y)$ is a set of lines parallel to the x-axis at the points of discontinuity of $H^{(k)}(y)$. If $l > r$, then the lines are more dense in the region $0 \leq y \leq r$, and if $l < r$, the lines are more dense in the region $r < y \leq 1$. As k approaches infinity, more and more of these lines appear in the unit interval in the y-direction, and the cumulative function $H^{(k)}(y)$

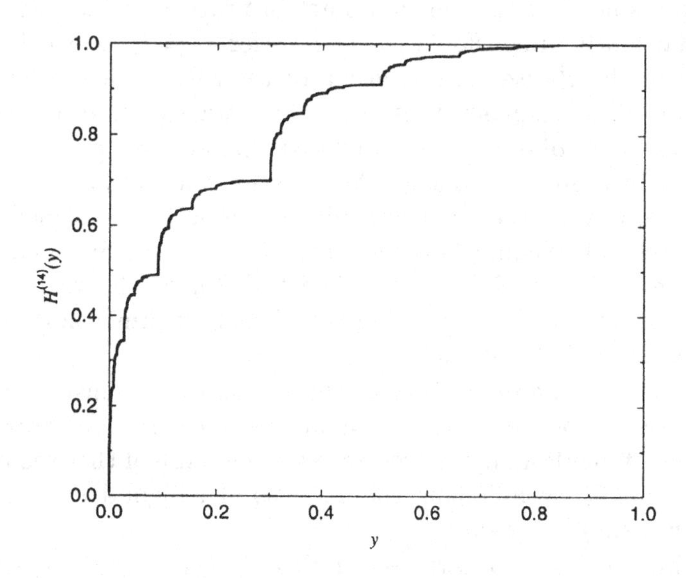

Fig. 13.5 The approximate Lebesgue singular function after fourteen iterations, $H^{(14)}(y)$. Here, $r = 0.3$.

approaches the Lebesgue singular function described above. This then provides a picture of the structure of the attractor.

The support of the attractor – the closure of the set of points in the attractor – is the unit square which, of course, has the box-counting dimension of 2. However, as mentioned above, we are interested in the dimension of the set which is the support of the SRB measure. For a two-dimensional invariant ergodic measure such as our SRB measure, Young has proved that the Hausdorff dimension, d_{H}, of this set is given by

$$d_{\mathrm{H}} = 1 + \frac{\lambda_{+}}{|\lambda_{-}|}. \tag{13.25}$$

For our case,

$$d_{\mathrm{H}} = 1 + \left| \frac{l \ln l + r \ln r}{l \ln r + r \ln l} \right| \leq 2, \tag{13.26}$$

with the equality sign holding only in the case that $r = l = \frac{1}{2}$. We have the result that the phase-space contraction produced by this map is not contraction to a point but rather to the attractor with Hausdorff dimension less than 2 (for $r \neq l$). We call this dimension the Hausdorff dimension of the SRB measure for the attractor. This is a general result for the attractors that appear in the phase-space of a thermostatted system. For a system with two non-zero Lyapunov exponents, Young's formula, (13.25), can be used to determine the Hausdorff dimension of the SRB measure. For higher-dimensional systems, there is a similar formula, still unproven but very likely true, called the *Kaplan–Yorke formula* which can be used to determine the Hausdorff dimension of the SRB measure of the attractor.

Although a careful analysis of the Hausdorff dimension of the SRB measure for the attractor we are considering would take us into some difficult analysis, we can get some sense of the reasoning behind (13.25) by following the analysis of Alligood *et al.* for a somewhat simpler case.

We imagine that we are considering a linear two-dimensional map, M', on the unit square such that in one direction the map has slope $\exp[\lambda_+]$, and in a perpendicular direction, the map has slope $\exp[\lambda_-]$. This is simpler than our model because in our case the slopes depend on the values of x. Now imagine that we follow the evolution under the map M' of a small square of length η, where $\eta \ll 1$, on a side in the unit square. After n iterations of the map, this small square will be stretched into a long thin rectangle of length $\eta \exp[n\lambda_+]$, and width $\eta \exp[n\lambda_-]$. If $n \gg 1$, the rectangle will be longer on a side than the side of the unit square, and it will be folded back into the unit square a large number of times. Now let's calculate the box-counting dimension of this transformed square. We need the definition of the previous discussion of the box-counting dimension d given in Chapter 11, i.e.,

$$d = \lim_{\epsilon \to 0} \frac{\ln N(\epsilon)}{\ln 1/\epsilon}, \tag{13.27}$$

where $N(\epsilon)$ is the number of boxes of length ϵ on a side needed to cover the set. Here, we can easily see that if we take $\epsilon = \eta \exp[n\lambda_-]$, then $N(\epsilon) = \exp[n(\lambda_+ - \lambda_-)]$. From this, it follows immediately that this box-counting dimension, d, of the fractal set produced by map M' is equal to the Hausdorff dimension d_{H} of the SRB measure given by (13.25). A careful analysis of our map M shows that Young's formula gives the Hausdorff dimension of the SRB measure of the fractal attractor, which is less than 2, while its box-counting dimension is indeed 2.

We now have a reasonably clear picture of the attractor generated by the map M. We have constructed an SRB measure of the attractor, and have developed an understanding of the type of phase-space contraction that is taking place in this system. It is now an appropriate time to consider the consequences of the map being reversible. If we use the Frobenius–Perron equation to construct an invariant SRB measure for the map M^{-1} instead of the map M, we find that an attractor is formed which is smooth in the y-direction and fractal in the x-direction. The SRB measure for this attractor, $\bar{\mu}_{\mathrm{SRB}}$, is determined by the cumulative function $\bar{G}(x,y) = (1-y)\bar{H}(x)$, where $\bar{H}(x)$ satisfies

$$\bar{H}(x) = \begin{cases} l + r\bar{H}(x/l) & \text{for } 0 \le x \le l; \\ l\bar{H}((x-l)/r) & \text{for } l < x \le 1. \end{cases} \tag{13.28}$$

Here we have taken the cumulative function to be defined by an integration over x and y, starting from the point $(1,1)$ instead of the origin. The reason for this is to take advantage of the time-reversal invariance of the map, for one can easily show that

$$\bar{G}(x,y) = (1-y)H(1-x), \tag{13.29}$$

where $H(x)$ is the function defined by (13.20). Therefore, using the time-reversal operator Φ, we note that $\bar{G} \cdot \Phi = G$. We will use this result in the proof of a fluctuation theorem later in this chapter.

The Lyapunov exponents of the map M^{-1} are identical to those for M, and the Hausdorff dimension of the SRB measure for the attractor of the reversed map is therefore the same as that for M. We can think of this attractor for map M^{-1} as a 'repeller' for the map M in the following way. Suppose we have a set of points which are distributed according to the SRB measure, $\bar{\mu}_{\mathrm{SRB}}$, on the attractor of M^{-1}. Then, since this measure is invariant, it will remain unchanged by either M or M^{-1}. However, suppose the distribution of this set of points differs ever so slightly from $\bar{\mu}_{\mathrm{SRB}}$. Then under the action of the map M the distribution of this set of points will eventually head to the distribution associated with the SRB measure, μ_{SRB}, for M. Now, the attractor for M^{-1} effectively functions as the repeller for the attractor of M (and vice versa), so we can apply the escape-rate formula to get an expression for the escape rate γ in terms of the Lyapunov exponent and KS entropy on the repeller. Remember that a repeller is an invariant set in phase-space but points arbitrarily close to this set move away from it in the course of time. The crucial point in our argument is that the SRB measure for the attractor for M^{-1}, $\bar{\mu}_{\mathrm{SRB}}$, is invariant under M. Therefore it is possible to think of $\bar{\mu}_{\mathrm{SRB}}$ as the invariant measure (under M) on the repeller. Using this invariant measure, we find that the Lyapunov exponents are given by

$$\begin{aligned} \lambda_+(\bar{\mu}_{\mathrm{SRB}}, M) &= [1 - \bar{H}(x = l)] \ln l^{-1} + \bar{H}(x = l) \ln r^{-1} \\ &= r \ln l^{-1} + l \ln r^{-1}, \quad \text{and} \qquad (13.30) \\ \lambda_-(\bar{\mu}_{\mathrm{SRB}}, M) &= r \ln r + l \ln l, \qquad (13.31) \end{aligned}$$

where $\bar{H}(x = l) = r$ is the cumulative distribution on the repeller for the interval $0 \le x \le l, 0 \le y \le 1$. We identify the positive Lyapunov exponent $\lambda_+(\bar{\mu}_{\mathrm{SRB}}, M)$ with the positive Lyapunov exponent on the repeller in the escape-rate formula. The KS entropy on the repeller, $h_{\mathrm{KS}}(\bar{\mu}_{\mathrm{SRB}}, M)$, can be computed using the method of Markov partitions described in Chapter 9 (see Exercise 13.1), and it is found to be

$$h_{\mathrm{KS}}(\bar{\mu}_{\mathrm{SRB}}, M) = l \ln l^{-1} + r \ln r^{-1}. \qquad (13.32)$$

Combining the results for the positive Lyapunov entropy and the KS entropy on the repeller, using the escape-rate formula, we obtain

$$\gamma = \lambda_+(\bar{\mu}_{\mathrm{SRB}}, M) - h_{\mathrm{KS}}(\bar{\mu}_{\mathrm{SRB}}, M) = (l - r) \ln \frac{l}{r}. \qquad (13.33)$$

Thus the escape rate from the repeller is minus the phase-space contraction rate, or minus the sum of the Lyapunov exponents for the map M. Of course, the attractor for M is also the repeller for M^{-1}, and the argument for the escape rate for this repeller is simply the time-reversal of the one given above.

13.3 The SRB theorem

It is now an appropriate time to generalize the description of an attractor given above to a general statement for the existence and properties of SRB measures for *hyperbolic* attractors. The maps

M and M^{-1} discussed above are examples of hyperbolic maps and their attractors are hyperbolic attractors. We recall from our previous discussion that a map M is hyperbolic if:

1. It has an invariant set Λ such that at every point x in Λ one can find stable and unstable invariant manifolds, $W_s(x)$ and $W_u(x)$, characterized by Lyapunov exponents strictly bounded away from zero by some constant non-zero value. Invariance of the manifolds means that $W_{s,u}(M(x)) = M(W_{s,u}(x))$, which is certainly true of the baker's map and the contracting map just considered.
2. The stable and unstable manifolds change continuously with x, such that as x changes continuously in Λ, then the unstable and stable manifolds vary continuously as well. For example, unstable directions do not suddenly turn into stable directions.
3. The stable and unstable manifolds intersect transversely, such that the angle of their intersection is strictly bounded away from zero. This condition rules out tangencies of the manifolds which leads to many serious complications in the analysis.

For hyperbolic attractors, Sinai, Ruelle and Bowen have proved a remarkable theorem which is a very strong generalization of Birkhoff's ergodic theorem.

(SRB Theorem) *Let M be a map which maps a region of phase-space, Γ, onto itself. Suppose M has a hyperbolic attractor Λ. Then there exists a unique (normalized) SRB measure μ_{SRB} with support on Λ with the property that:*

1. If U is a neighborhood of Λ such that $M^n(U) \subset U$ for all $n > 0$, and $\Lambda = \cap_{n \geq 0} M^n(U)$; and
2. if f is a continuous function on U;

then, for almost all $x \in U$ (with respect to Lebesgue measure),

$$\lim_{N\to\infty} \frac{1}{N} \sum_{n=0}^{N-1} f(M^n(x)) = \int_\Lambda f(x)\mu_{\mathrm{SRB}}(dx). \tag{13.34}$$

In the example discussed above, the set U is the unit square, the map M is given by (13.1), and μ_{SRB} is given by (13.23). The SRB theorem applies to all similar hyperbolic attractors. Of course, the calculation of the SRB measure is not as simple in most cases as the example given here. In general, only approximate calculations of these measures are available. Periodic orbit expansions are often used to compute these approximate SRB measures, and we discuss these expansions in Chapter 15.

13.4 Entropy in terms of SRB measures

Now that we have some understanding of SRB measures we can turn to the resolution of the sign difficulty in the rate of entropy production that we encountered in the treatment of thermostatted systems in the previous chapter. Of course, the crucial point is that the distribution function is not a differentiable function in the steady state, so that the manipulations made in that discussion really had no meaning. To do things properly, one must base an expression for the entropy and rate of entropy production on the proper SRB measures for the sets in phase-space. The methods for doing this were first proposed by Gaspard and by Tél and co-workers. Here we follow an extension of this work by Gilbert and Dorfman, and we refer the reader to this paper for more details.

We begin by supposing that we have a map, M, and a phase space, Γ. We also suppose that we can make a fine Markov partition of Γ into rectangles, which we denote by A_i, and the collection of rectangles is denoted by $\mathcal{A}$. The SRB measure of a rectangle at time t, say, is denoted by $\mu(A_i, t)$, and the phase-space volume of A_i is denoted by $\nu(A_i)$. Then we define the entropy at time t of a subset B of Γ, with respect to this partition, as

$$S(B, t|\mathcal{A}) = \sum_{A_i \in B} \mu(A_i, t) \left[\frac{\mu(A_i, t)}{\nu(A_i)} - 1 \right], \tag{13.35}$$

and we suppose that the Markov partition is sufficiently fine that the set B can be considered to be the union of a large number of the rectangles A_i. This definition of the entropy is a natural one if one replaces $\rho(\Gamma, t)$ by $\mu(A_i, t)/\nu(A_i)$ and $\rho(\Gamma, t)d\Gamma$ by $\mu(A_i)$ in the standard definition of the Gibbs entropy in statistical mechanics, $\int d\Gamma \rho \ln \rho$, and replaces the integral by a sum. One might say that the need to use a singular SRB measure for a thermostatted system requires us to use a *coarse-grained entropy*. The entropy defined by (13.35) is a finite quantity, although one can show that in the limit where the partition becomes arbitrarily fine, this entropy can take the value of minus infinity. An example is the entropy of the map discussed in the previous section (see Exercise 13.2).

We write the rate of change of the entropy with time as the sum of three terms,

$$\begin{aligned}\Delta S(B,t|\mathcal{A}) &= S(B,t+1|\mathcal{A}) - S(B,t|\mathcal{A}) \\ &= \Delta_{\mathrm{e}} S(B,t|\mathcal{A}) + \Delta_{\mathrm{th}} S(B,t|\mathcal{A}) + \Delta_{\mathrm{i}} S(B,t|\mathcal{A}),\end{aligned} \tag{13.36}$$

where we have identified three sources for the change of entropy in set B. The first term, $\Delta_{\mathrm{e}} S(B,t|\mathcal{A})$, is the *net flow* of entropy into the set B in one time step due to points moving into and out of B, given by

$$\Delta_{\mathrm{e}} S(B,t|\mathcal{A}) = S(M^{-1}(B),t|\mathcal{A}) - S(B,t|\mathcal{A}). \tag{13.37}$$

The second term, $\Delta_{\mathrm{th}} S(b,t|\mathcal{A})$ is the change in the entropy of B due to the presence of the thermostat. Here, we identify the effects of the thermostat by the change in the volume of the sets under their time evolution. That is, we define this change in entropy by

$$\begin{aligned}\Delta_{\mathrm{th}} S(B,t|\mathcal{A}) &= S(B,t+1|\mathcal{A}) - S(M^{-1}(B),t|M^{-1}\mathcal{A}) \\ &= -\sum_{A_i \in B} \mu(A_i,t) \ln \frac{\nu(M^{-1}(A_i)}{\nu(A_i)},\end{aligned} \tag{13.38}$$

where we define the entropy change due to the thermostat as the difference between the entropy of B at time $t+1$ and the entropy of the pre-image of B at time t, where the entropy of the pre-image sets of B are calculated using the pre-images of the sets in the partition $\mathcal{A}$. If the map is volume-preserving, then the entropy change in B due to the thermostat vanishes. By combining (13.37)–(13.38) we can obtain an expression for the irreversible production of entropy $\Delta_{\mathrm{i}} S(B,t|\mathcal{A})$ as

$$\Delta_{\mathrm{i}} S(B,t|\mathcal{A}) = S(M^{-1}B,t|M^{-1}\mathcal{A}) - S(M^{-1}B,t|\mathcal{A}). \tag{13.39}$$

This expression for the irreversible entropy production in the set B is the difference in entropy of the pre-image sets at two levels of resolution. A positive entropy production then results from the fact that information is being lost if the resolution of the sets is becoming coarser in a critical direction. The pre-image sets of the A_i are coarser along the stable directions than the sets of A_i, and it is along the stable directions that the SRB measures are very irregular. Thus, a coarser partition will miss some of the irregular structure and information will be lost! Positive irreversible entropy production then is due to the loss of information on some fine scales as the system's dynamics produces irregular densities on finer and finer scales. Now we can define an intrinsic entropy change of the set B, $\Delta_{\text{in}}S(B,t|\mathcal{A})$, by removing the term due to the thermostat, as

$$\begin{aligned}\Delta_{\text{in}}S(B,t|\mathcal{A}) &= \Delta S(B,t|\mathcal{A}) - \Delta_{th}S(B,t|\mathcal{A}) \\ &= S(M^{-1}B,t|M^{-1}\mathcal{A}) - S(B,t|\mathcal{A}). \quad (13.40)\end{aligned}$$

Finally, we consider what happens in a nonequilibrium steady state. In such a state, $\mu(A_i, t+1) = \mu(A_i, t)$, and it then follows that $\Delta S(B,t|\mathcal{A}) = 0$. The overall entropy change is zero. Moreover, it also follows from the steady-state Frobenius–Perron equation that $\mu(A_i) = \mu(M^{-1}A_i)$, so the measure of a set is invariant. Since the overall entropy change of B is zero, the intrinsic entropy change of B is equal to the rate of flow from the system to the thermostat, that is, $-\Delta_{\text{th}}S(B|\mathcal{A})$, which is positive if there is a contraction of the system onto an attractor. A little bit of analysis shows (see Exercise 13.3) that in a steady state, one can take the limit of infinitesimally fine Markov partitions to show that in this limit, the intrinsic entropy production becomes

$$\Delta_{\text{in}}S(B) = \int_{M^{-1}(B)} \mu(d\Gamma) \ln \frac{1}{J(\Gamma)} \quad (13.41)$$

where $J(\Gamma)$ is the Jacobian of the map M, $J(\Gamma) = |dM(\Gamma)/d\Gamma|$. In the case that B corresponds to the full phase-space, the intrinsic entropy change is simply the result needed in the previous chapter, namely,

$$\int_{M^{-1}(B)} \mu(d\Gamma) \ln \frac{1}{J(\Gamma)} = -\sum_i \lambda_i. \quad (13.42)$$

Thus, a careful treatment of the entropy production in a nonequilibrium steady state leads to the relation that the intrinsic entropy change of the phase-space is the negative of the entropy flow from the thermostat to the system, and is the negative of the sum of all of the Lyapunov exponents of the system.

It is important to mention that this discussion is not restricted to systems with phase-space contraction. In fact, Gaspard's arguments were first applied to a nonequilibrium steady state produced by purely Hamiltonian dynamics. He considered a system with (infinite) particle reservoirs placed in such a way that a steady density gradient was established and maintained across the system. One can imagine, for example, that there is a low-density reservoir placed at $x = -L/2$, and a high-density reservoir at $x = L/2$, such at over the interval $-L/2 \leq x \leq L/2$, ordinary Hamiltonian dynamics governs the motion of the particles. Such models can be easily constructed using baker's maps, or Lorentz gases. Gaspard and Tasaki have shown that a very complicated fractal structure develops in such Hamiltonian, nonequilibrium steady states, and Gaspard then used the ideas described above to show that a careful analysis of the appropriate measures for this steady state leads to a positive entropy production in the system. For further details the reader should consult the papers of Tasaki and Gaspard listed at the end of the chapter.

13.5 The Gallavotti–Cohen fluctuation formula

As a result of the above analysis of the entropy production in thermostatted systems, we can now give an example of one of the interesting results obtained in the analysis of fluid systems using ideas from dynamical systems theory. This is the fluctuation formula of Gallavotti and Cohen. This fluctuation formula is based on the observation of the motions of fluid particles in thermostatted systems undergoing shear flow. It was found by Evans, Morriss, and Cohen that over finite time intervals, there was a small but non-zero probability of observing motions that were not in accord with the macroscopic laws of irreversible thermodynamics. For example, in a thermostatted Lorentz gas with a charged particle in an electric field, one might observe the particle moving in a direction opposite to the field. On a long time-scale, the particles will, on the average, move in the direction of the field, but there will be occasional fluctuations where the motion is opposite to the field. The question is to express the probability that such motions will be observed over a time interval, τ, say.

This problem is attacked by associating a finite-time entropy production, ϵ_τ, with a trajectory of length τ, and then comparing the probability that ϵ_τ will have some value a, say, with the probability that ϵ_τ will have value $-a$. The Gallavotti–Cohen fluctuation formula is

$$\frac{\text{Prob}(\epsilon_\tau = a)}{\text{Prob}(\epsilon_\tau = -a)} = e^{a\tau}. \tag{13.43}$$

As τ gets large, the ratio of these probabilities goes to zero or infinity depending on the sign of a.

This result was proved for reversible dynamical systems using SRB measures and Markov partitions. Here, we will give an example of this result using the SRB measure associated with the map M defined by (13.1). We define the finite-time entropy production at a phase point X by

$$\epsilon_\tau(X) = \frac{1}{\tau}\sum_{j=0}^{j=\tau-1} \ln \frac{1}{J(M^j(X))}, \tag{13.44}$$

where $J(M^jX)$ is the Jacobian $|dM^{j+1}(X)/dM^j(X)|$. Now we identify the ratio of the probabilities in (13.43) with the ratio of the SRB measures for the regions where $\epsilon_\tau(X) = a$ and $\epsilon_\tau(X) = -a$, as

$$\frac{\text{Prob}(\epsilon_\tau = a)}{\text{Prob}(\epsilon_\tau = -a)} = \frac{\mu_{\text{SRB}}(X|\epsilon_\tau(X) = a)}{\mu_{\text{SRB}}(X|\epsilon_\tau(X) = -a)}. \tag{13.45}$$

It is now an easy matter to compute the right-hand side of (13.45). First, we take advantage of the time-reversibility of the map M to note that the SRB measure of the set of points, X, where $\epsilon_\tau(X) = -a$, is equal to the SRB measure defined with respect to M^{-1}, denoted earlier in this chapter by $\bar{\mu}_{\text{SRB}}$, of the set of points, X, where $\epsilon_\tau(X) = a$. There are a number of ways to see this for our map, and certainly one way is to use the fact that a time-reversal of a trajectory changes the sign of ϵ_τ, and that the time-reversal of μ_{SRB} is $\bar{\mu}_{\text{SRB}}$, as follows from (13.29). Now for any X we can write

$$\epsilon_\tau(X) = \frac{\tau_1}{\tau}\ln\frac{r}{l} + \frac{\tau_2}{\tau}\ln\frac{l}{r}, \tag{13.46}$$

where τ_1 is the number of times the iterates of X are in the interval $0 \le x < l$, and $\tau_2 = \tau - \tau_1$ is the number of times the iterates are in the interval $l \le x < 1$. The SRB measure of the region on the unit square where τ_1 and τ_2 have specified values is

$$\mu_{\text{SRB}}(\tau_1, \tau - \tau_1) = l^{\tau_1} r^{\tau_2}. \tag{13.47}$$

Similarly,

$$\bar{\mu}_{\text{SRB}}(\tau_1, \tau_2) = r^{\tau_1} l^{\tau_2}. \tag{13.48}$$

Consequently,

$$\frac{\text{Prob}(\epsilon_\tau = a)}{\text{Prob}(\epsilon_\tau = -a)} = \left(\frac{l}{r}\right)^{\tau_1}\left(\frac{r}{l}\right)^{\tau_2} = e^{\tau\epsilon_\tau} = e^{a\tau}, \tag{13.49}$$

if we choose τ_1 and τ_2 so that $\epsilon_\tau = a$. This, then, is an example of the fluctuation formula and it is worth noting that we have demonstrated it for a map not considered by Cohen and Gallavotti, since their arguments used properties of smooth Anosov flows or of diffeomorphisms, while the map used here is not in these classes. However, the key elements are the time-reversibility of the map and the relation between the SRB measures for the map and its inverse.

13.6 Gibbs measures

In our first discussion of repellers in Chapter 12, we introduced the notion of a topological pressure from which we could compute all of the dynamical properties of trajectories on the repeller. These properties included the escape rate, the KS entropy, and the positive Lyapunov exponent for the one-dimensional map studied in that chapter. There we introduced the thermodynamic formalism which looks very much like the formalism of equilibrium statistical mechanics. Here, we are going to look somewhat deeper into this analogy between statistical thermodynamics and chaos theory in order to introduce some new quantities such as the topological entropy and to illustrate the utility of Gibbs measures in a broader context. In particular, we want to make the important connection between the dynamical measure on phase-space trajectories and the microcanonical ensemble of statistical thermodynamics.

We begin by returning to the definition of the dynamical partition function given by (11.28) for a one-dimensional map with escape,

$$Z_n(\beta) = \sum_{i=1}^{2^n}(l_i)^\beta, \tag{13.50}$$

where n is the number of iterations of the map, β is a parameter, and l_i is the length of the ith interval of initial points that remain in the unit interval after n iterations of the map. The point that we wish to stress here is that l_i is determined by the history of the points that start in this interval, that is, by the number, n', of times that points were acted upon by the part of the map with stretching factor $1/p_0$, and by the number, $n - n'$, of times, acted upon by the part with stretching factor $1/p_1$, in a total of n steps. That is, $l_i = p_0^{n'} p_1^{(n-n')}$. This result can be put in a more suggestive way which reveals the idea underlying the construction of Gibbs measures. To do this, we observe that we can write, for every point x in the interval labeled by the subscript i,

$$\begin{aligned} l_i &= p_0^{n'} p_1^{(n-n')} \\ &= \exp\left[-\sum_{j=0}^{n-1} \ln |M'(x_j)|\right] \\ &= \exp n \left[-\frac{1}{n}\sum_{j=0}^{n-1} \ln |M'(M^j(x))|\right], \end{aligned} \tag{13.51}$$

where M is the map defined in Sec. 11.1. Next we note that $M'(x_j)$ is the slope of the map evaluated at the point $x_j = M^j(x)$. When n is large, the intervals become quite small, and we can take x to be a typical point in the small interval under discussion. We note from (13.51) that we can associate a 'stretching factor' $u(n, x)$ with this interval, where

$$u(n, x) = \frac{1}{n}\sum_{j=0}^{n-1} \ln |M'(x_j)|, \tag{13.52}$$

so that (13.51) becomes

$$l_i = e^{-nu(n,x)}. \tag{13.53}$$

Note that $u(n, x)$ may be described as a finite-time Lyapunov exponent for the point x. Similarly, the dynamical partition function, $Z_n(\beta)$, can be written as

$$Z_n(\beta) = \sum_{\{x_i\}} e^{-n\beta u(n,x_i)} \tag{13.54}$$

where the set, $\{x_i\}$, of points is chosen in the following way: We suppose that $n \gg 1$, so that the n-step approximation to the fractal repeller is covered by a large number, 2^n, of disjoint intervals labeled by index i. In each interval i, pick a representative point, and collect these points, one from each interval, as the set $\{x_i\}$. Thus we have expressed the dynamical partition function in terms of a set of representative points, one from each member of a set of intervals which together cover this approximate repeller, and in terms of the stretching factor associated with each of these points.

Now, in calculating the dynamical quantities such as the escape rate, the KS entropy, and so on, we need to take derivatives of the logarithm of the partition function. This will introduce a set of 'Boltzmann–Gibbs' weights, $w_i(n, \beta)$, for each of the intervals, given by

$$w_i(n, \beta) = \frac{1}{Z_n(\beta)} e^{-n\beta u(n, x_i)}, \tag{13.55}$$

an expression which looks remarkably like the Boltzmann–Gibbs weights familiar from statistical thermodynamics. It is important to mention that in the special case of $\beta = 0$, or 'infinite temperature', the dynamical partition becomes simply the number of disjoint sets needed to cover the n-step approximation to the repeller. Let us suppose that the minimum number of such sets is $\mathcal{N}(n)$, (in our case $\mathcal{N}(n) = 2^n$), then we define a *topological entropy*, h_{top}, by the relation

$$h_{\text{top}} = \lim_{n \to \infty} \frac{1}{n} \ln \mathcal{N}(n) = \lim_{n \to \infty} \frac{1}{n} \ln Z_n(\beta = 0). \tag{13.56}$$

Clearly, in our case, $h_{\text{top}} = \ln 2$, and, in general, the topological entropy measures the rate of growth with time of the number of disjoint sets needed to cover all of the initial points whose images still remain in the indicated region after a given time. If this rate of growth is exponential, then the topological entropy will be non-zero.

Those familiar with the derivation of the Boltzmann–Gibbs distributions in statistical thermodynamics know that these distributions are usually obtained through some variational principle. That is, they are constructed by maximizing an entropy function of an appropriate type (depending on the possibility of exchanging particles and/or energy with a reservoir). It is both reassuring and useful to know that the Boltzmann–Gibbs distribution that we have just constructed above can also be derived from a similar variational procedure. One defines an entropy functional, usually the topological pressure, which corresponds to the negative of the thermodynamic Helmholtz free energy, and then finds a distribution which maximizes the pressure.

This construction of Boltzmann–Gibbs weights for phase-space regions, usually referred to as *Gibbs measures*, can be generalized in a number of ways. We can

1. consider repellers in phase-spaces of higher dimension;
2. conceivably use functions other than $\ln|M'(x)|$;
3. make the parameter β into a multi-dimensional object so as to include a number of different functions in the definition of a generalized dynamical partition function;
4. extend the types of systems for which a dynamical partition function can be defined to include Markov processes.†
5. generalize the definition of the dynamical partition function to a broader class of sets than just repellers; in fact, we can include the entire phase-space available to the system, if we want to.

Here, we consider the first and last of these possibilities, and refer the reader to the literature for a discussion of the others.

We begin by considering the dynamical partition function for a *repeller* in some higher-dimensional phase-space. We suppose that the dynamics is volume-preserving or symplectic, so that if the dynamics is indeed chaotic, the sum of *all* of the Lyapunov exponents is equal to zero. Suppose further that we are considering an open system and we want to identify the dynamical properties of the set of points which always remain inside some specified boundaries placed in phase-space. These properties are needed for an application of the escape-rate formalism to the calculation of transport properties, as discussed in Chapter 12, for example. We consider an Anosov system so that we can identify stable and unstable directions at every point in phase-space. Suppose now that we sprinkle a very large number of points uniformly, with respect to Lebesgue measure, over the phase-space. As in the one-dimensional case, the set of initial points whose trajectories remain within the specified boundaries in phase-space over a time interval $[0, N]$ will be called the approximate repeller. If N is large enough, this approximate repeller will be composed of a large number of small sets, which we label with index i. A typical point, $\mathbf{x}_i$, is selected in each of these sets, and we form a set of typical points,

† This makes it easily possible to define various entropies and other dynamical properties for stochastic systems. An example is descibed in Chapter 16.

denoted as $\{\mathbf{x}_i\}$. To determine the weights associated with each of these sets, we note that the escape from the specified region will be determined by the *unstable* directions in phase-space and with the stretching factors associated with each of them. The initial set of points in the phase-space will be compressed in the stable directions while they will be expanded in the unstable directions. This expansion is responsible for the escape, as we saw in the one-dimensional case. Therefore in a higher-dimensional setting, we assign weights $w_i(N, \beta)$ to each of the sets in the approximate repeller, based upon the stretching factors in the unstable directions, by

$$w_i(N, \beta) = \frac{1}{Z_N(\beta)} \exp\left[-\beta N \sum_{\alpha=1}^{d_\mathrm{u}} u_\alpha(N, \mathbf{x}_i)\right], \tag{13.57}$$

where the dynamical partition function $Z_N(\beta)$ is given by

$$Z_N(\beta) = \sum_{\{\mathbf{x}_i\}} \exp\left[-\beta N \sum_{\alpha=1}^{d_\mathrm{u}} u_\alpha(N, \mathbf{x}_i)\right]. \tag{13.58}$$

The quantities $u_\alpha(N, \mathbf{x}_i)$ are the stretching factors in each of the expanding directions, which we label by the subscript α, which ranges from 1 to d_u, the total number of expanding directions. These stretching factors can be defined readily for hyperbolic maps, but the definition requires some extended analysis and is presented by Gaspard and Dorfman. Here, we give the definition in the special case that all of the expanding and contracting directions are mutually orthogonal with respect to a Euclidean metric in the phase-space. These circumstances arise naturally in baker's maps and in many toral automorphisms such as the Arnold cat map discussed earlier. In this case,

$$u_\alpha(N, \mathbf{x}_i) = \frac{1}{N} \sum_{n=0}^{N-1} \ln |M'_\alpha(\mathbf{M}^n(\mathbf{x}_i))|. \tag{13.59}$$

Here, $\mathbf{M}$ is the full map in the phase-space, M_α is its component in the direction labeled by α, and the prime denotes the derivative with respect to the coordinate in the α direction. The topological pressure, $P(\beta)$, is then defined as

$$P(\beta) = \lim_{N\to\infty} \frac{1}{N} \ln Z_N(\beta). \tag{13.60}$$

The dynamical quantities on the repeller can now be defined in terms of $P(\beta)$ as

$$P(1) = -\gamma, \tag{13.61}$$

$$\left.\frac{dP(\beta)}{d\beta}\right|_{\beta=1} = \sum_{\lambda_i>0} \lambda_i, \tag{13.62}$$

$$h_{\mathrm{top}} = P(\beta = 0), \tag{13.63}$$

$$h_{\mathrm{KS}} = \left.\frac{dP(\beta)}{d\beta}\right|_{\beta=1} + P(1), \tag{13.64}$$

$$P(\beta_c) = 0. \tag{13.65}$$

Here, β_c is the value of β, with $0 < \beta_c < 1$, where the pressure function vanishes for an open system. It can be shown that β_c is the fractal dimension of the intersection of an arbitrary line with the stable manifolds of the fractal repeller. This analysis is essentially formal, since in order to actually compute these quantities, we need to specify the exact structure of the bounded region in phase-space from which the points will be allowed to escape. This region determines the structure of the approximate as well as the full repeller and determines where the points $\mathbf{x}_i$ are to be found. Thus, even in a case where the stretching factors are trivial, as in the baker's map with holes placed somewhere in the unit square, the points $\mathbf{x}_i$ will depend upon the exact placement of the holes. As alluded to above, this analysis can be extended to general hyperbolic maps as well as to hyperbolic flows where the time is a continuous variable. For reversible maps and flows, we should be careful to construct the repeller in such a way that it is time-reversal invariant, but that is a point which we leave to more detailed discussions in the literature.

We now turn to the issue of defining Gibbs measures on the entire phase-space, not just a repeller. This issue was resolved by Bowen and by Walters who introduced the notion of a *separated set* for phase-space.† The idea is to decompose the phase-space into very small sets and then, taking a typical point in each of the small sets, use it to compute the Boltzmann–Gibbs weight of the small set. To do this, we suppose that some measure-preserving,

† The notion of separated sets allows for a convenient definition of the topological entropy of a dynamical system. Katok showed how separated sets can be used to define the KS entropy as well (see *Further reading*).

reversible map, $\mathbf{M}$, is acting on our phase-space, and we use it to define an (ϵ, N)-separated set as follows. Consider the time N, and define a new distance, $\rho_N(\mathbf{x}_1, \mathbf{x}_2)$, between two phase points, $\mathbf{x}_1$ and $\mathbf{x}_2$, as

$$\rho_N(\mathbf{x}_1, \mathbf{x}_2) = \max_{-N \le n \le N} \|\mathbf{M}^n(\mathbf{x}_1) - \mathbf{M}^n(\mathbf{x}_2)\|, \tag{13.66}$$

where $\|\mathbf{x}_i - \mathbf{x}_j\|$ denotes some metric distance between points $\mathbf{x}_i$ and $\mathbf{x}_j$ in phase-space. If two points satisfy $\rho_N(\mathbf{x}_1, \mathbf{x}_2) < \epsilon$, then the trajectories of two phase points starting at $\mathbf{x}_1$ and $\mathbf{x}_2$ would remain within a distance ϵ over the time interval $[-N, N]$. We take our phase-space to be invariant under the map $\mathbf{M}$, and construct a set of points $\mathcal{S} = (\mathbf{Y}_1, \ldots, \mathbf{Y}_S)$ such that $\rho_N(\mathbf{Y}_i, \mathbf{Y}_j) > \epsilon$ for all $i \neq j$. If our phase-space is compact, one can always find a set $\mathcal{S}$ with a finite number of points. For example, one can cover the phase-space with a collection of sets with ρ_N radius of $2\epsilon'$. Since the space is compact, we can cover it with a finite subset of these sets, and we can take the $\mathbf{Y}_i$ to be the centers of each of the sets in this finite collection. Since there are only a finite number of these points, there will be some value of ϵ for which the set of the $\mathbf{Y}_i$ will be an (ϵ, N)-separated set.

The set $\mathcal{S}$ is called a (ϵ, N)-separated set for the phase-space. Since (ϵ, N)-separated sets exist with finite numbers of points, there is at least one with some maximum number of points, which we denote by $\mathcal{S}_N$. For a hyperbolic system the trajectories separate exponentially with time, the number of points in $\mathcal{S}_N$ will increase exponentially with N because the points become separated by smaller and smaller distances as measured by the usual metric on the phase-space, as N becomes large. Again, the number of points as well as the ρ_N distance between them is determined by the rates of separation of points in the unstable directions in the phase-space for the forward and backward motion in time.

We now define a partition function for the phase-space by

$$Z_N(\beta, \epsilon) = \sum_{\mathbf{Y}_i \in \mathcal{S}_N} \exp\left[-2N\beta u(N, \mathbf{Y})\right], \tag{13.67}$$

where $u(N, \mathbf{Y})$ is the total (i.e., including all of the expanding directions) stretching factor over the time interval $[-N, N]$ for a trajectory that is initially at $\mathbf{Y}$. The topological pressure is then given by

$$P(\beta) = \lim_{\epsilon\to 0}\lim_{N\to\infty}\frac{1}{2N}\ln Z_N(\beta,\epsilon). \tag{13.68}$$

An immediate consequence of this definition is the fact that the topological entropy measures the rate of the exponential growth of the number of points in S_N. The topological pressure must also satisfy the condition that

$$P(\beta = 1) = 0, \tag{13.69}$$

since there is no escape from the full phase-space. Another consequence of this construction is the Gibbs measure of the set that contains the point $\mathbf{Y}_i$, given by

$$w_i(\beta, N, \epsilon) = \frac{1}{Z_N(\beta,\epsilon)}\exp\left[-2N\beta u(N,\mathbf{Y}_i)\right]. \tag{13.70}$$

In Fig. 13.6 we illustrate the general features of the topological pressure as a function of the parameter β for both closed and open hyperbolic systems.

This is all very formal, so it is instructive to consider a simple model. Let us apply the above method to compute the dynamical partition function for the baker's map on the unit square. An (ϵ, N)-separated set is easily constructed by picking some small value for ϵ and then dividing the unit square up into a large number of small squares with sides parallel to the x- and y-axes and of length $\epsilon'/\exp[N\lambda_+] = \epsilon'/2^N$, where ϵ' is just a bit larger than ϵ, say 1.01ϵ. If we take the midpoints of each of these small squares to be the points $\mathbf{Y}_i$, then it is easy to see that the set of the $\mathbf{Y}_i$ satisfies the condition for an (ϵ, N)-separated set, since the maximum separation of any two different $\mathbf{Y}_i$s over the interval $[-N, N]$ will be at least ϵ'. Moreover, the number $\mathcal{N}(N)$ of such small squares will be equal to

$$\mathcal{N}(N) = \left[\epsilon'/2^N\right]^{-2}, \tag{13.71}$$

which leads immediately to the result that the topological entropy is $h_{\text{top}} = \ln 2$. The baker's map is so simple that one can easily see that the stretching factors for each of the $\mathbf{Y}_i$ is $\ln 2$, so that the Gibbs measure for each of the small squares and the dynamical partition function for this system are given by

$$w_i(\beta, N) = \frac{2^{-2N\beta}}{Z_N(\beta)}, \tag{13.72}$$

Closed

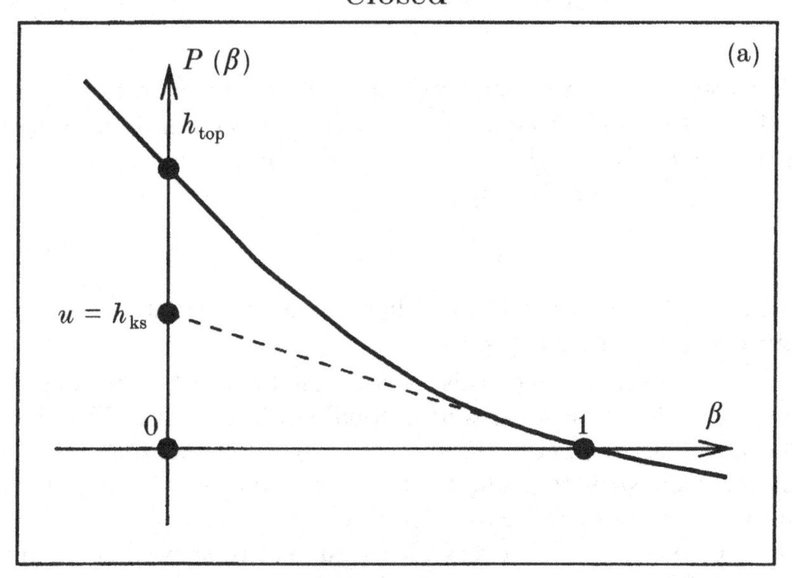

Open

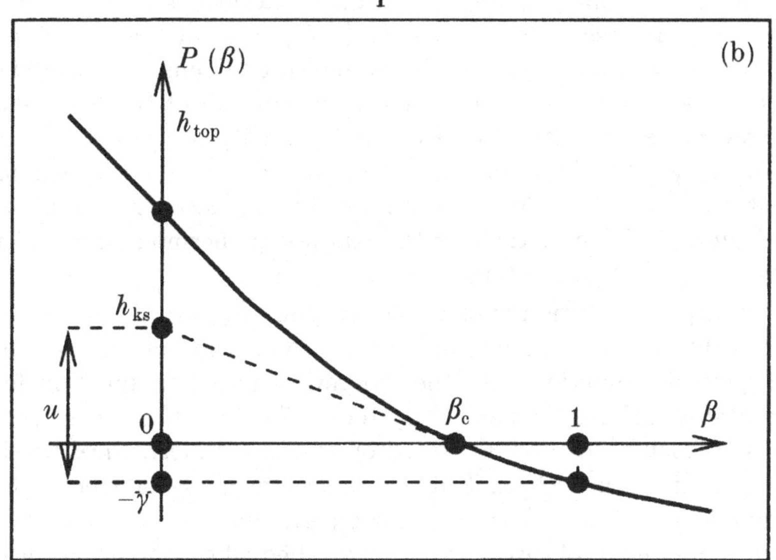

Fig. 13.6 Schematic illustration of the topological pressure and related dynamical quantities for (a) a closed hyperbolic system; and (b) an open hyperbolic system.

$$Z_N(\beta) = \epsilon'^2 \sum_i 2^{-2N\beta} = 2^{2N(1-\beta)}. \tag{13.73}$$

We have included a normalizing factor ϵ'^2 in the definition of the partition function which has no effect on the dynamical properties of the system, it simply makes the partition function neater. The topological pressure is simply

$$P(\beta) = (1 - \beta) \ln 2, \tag{13.74}$$

and the values of all of the familiar dynamical quantities are obtained easily from this result.

One obvious result of this calculation for the baker's map is that all of the small squares have equal Gibbs measure. The question arises as to whether this is a trivial result, holding only for simple maps with constant stretching factors, or whether this is an indication of some deeper result. The latter case is, in fact, the truth. One can present a very nice argument to show that the dynamical Gibbs measure of a small set on, say, a constant-energy surface in the phase-space for a mechanical system and the microcanonical ensemble measure $d\sigma/|\nabla H|$ coincide (up to trivial factors) when $\beta = 1$ in the Gibbs measure.† Thus, we have the result that the microcanonical measure (and all of the other ensemble measures that can be derived from it) coincide with the natural dynamical measure based upon the stretching of regions in the unstable directions. So, we are getting closer and closer to the dynamical foundations of the statistical thermodynamics, at least for hyperbolic systems.

Finally, we mention that for Ising-type magnetic systems one can make a one-to-one mapping of the calculation of the statistical partition function onto the dynamical partition function for a suitably defined dynamical system. The fact that there is a phase transition can be described by statistical mechanical partition functions suggests that phase transitions might be seen in the dynamical partition functions for some systems as a function of β. In fact, a number of cases have been studied where the topological pressure has some kinds of singularities at certain values of β. We refer the reader to the literature for further details.

† See the article by Gaspard and Dorfman [GD95].

13.7 Further reading

Much of the material on SRB measures is taken from a paper by Tasaki, Gilbert, and Dorfman [TGD98]. I am indebted to my co-authors S. Tasaki and T. Gilbert for many helpful conversations on this subject. The SRB theorem is carefully explained in the review paper of Ruelle and Eckmann [RE85]. More advanced discussions can be found in Katok and Hasselblatt [KH95]. The application of SRB measures to statistical physics is described in papers by Gallavotti and Cohen [GC95b], Cohen [Coh97c], Gallavotti [Gal95], Ruelle [Rue97a] and by Gallavotti and Ruelle [GR97]. Of course, the original papers by Sinai [Sin72], Ruelle [Rue76], and by Bowen and Ruelle [BR75], among others, should not be overlooked. Detailed discussions of the definition of the entropy and rate of entropy production in nonequilibrium steady states, based on SRB measures, can be found in the book and papers of Gaspard [Gas98, Gas97b], in the paper of Tasaki and Gaspard [TG95],in the papers of Vollmer, Tél, and Breymann [TVB96, BTV96, VTB97, BTV98, VTB98], and in Gilbert and Dorfman [GD98]. The Gallavotti–Cohen fluctuation formula was inspired by the paper of Evans, Cohen, and Morriss [ECM93], and a proof for smooth Anosov systems can be found in the paper of Gallavotti and Cohen [GC95b]; a less mathematical proof is discussed by Cohen [Coh97c]. A very nice presentation of this and related subjects can be found in the lecture notes of Ruelle [Rue98]. Further numerical studies of the fluctuation formula have been carried out by Bonetto, Gallavotti, and Garrido [BGG97]. Gallavotti has also discussed the connections between the fluctuation formula, the Green–Kubo formulae, and the Onsager reciprocal relations [Gal96]. See also the paper by Rondoni and Cohen for a related discussion [RC98]. Derivations of the Gallavotti–Cohen fluctuation theorem based upon a stochastic dynamics have been given by Kurchan [Kur98] and by Leobwitz and Spohn [LS98]. A general discussion of the fluctuation theorem and related topics, based upon Gibbs measures, has been provided by Maes [Mae98].

Gibbs measures are discussed in the papers by Sinai [Sin72], and in the books by Bowen [Bow75] and by Ruelle [Rue78]. More recent discussions can be found in the books by Falconer [Fal97], Beck and Schlögl [BS93], Gaspard [Gas98], and by Keller [Kel98].

The use of separated sets to define the topological entropy of dynamical systems is described in the book of Walters [Wal81], and the extension to the KS entropy is given by Katok [Kat82]. A very nice description of Gibbs and SRB measures and thermodynamic formalism has been given by Baladi [Bal99]. The papers by Bohr and Rand [BR87], Bedford [Bed91], and Gaspard and Dorfman [GD95] describe the construction of Gibbs measures for fractal repellers. Dynamical phase transitions are described in a variety of places including the books of Beck and Schlögl, and Gaspard, just cited.

Application of Gibbs, SRB, and related measures, to systems with escape are discussed in the papers by Pianigiani and Yorke [PY79] and by Chernov and Markarian [CM97a, CM97b], in addition to those listed above.

13.8 Exercises

1. Derive (13.28)–(13.31) for the SRB measure, the Lyapunov exponents and the KS entropy of the repeller described in Sec. 13.2.

2. Show that the entropy defined by (13.35) is minus infinity in the limit of infinitely fine resolution for the map M discussed in Sec. 13.1.

3. Derive (13.41) and (13.42).

14
Fractal forms in Green–Kubo relations

In the previous chapter, we discussed two possible approaches to the connection between microscopic chaotic dynamics and macroscopic transport phenomena. Here, we are going to discuss yet another approach, which shows a dramatic connection between the Green–Kubo formula for the coefficient of diffusion and a fractal structure that results from the evaluation of the velocity autocorrelation function for a simple, one-dimensional, periodic map. While the model we consider here is very special, not to say artificial, it has a number of features that are of more general interest. As noted in the reference list at the end of the chapter, the ideas described here are based upon the work of Gaspard and Tasaki. At the end of this chapter, we will indicate how one might find general features in this approach, as illustrated in the special case described here.

14.1 The Green–Kubo formula for maps

We consider a one-dimensional map, $x_{n+1} = M(x_n)$. Now we suppose this map is such that it operates over the interval $-L < x < L$, and, further, that there exists an invariant equilibrium distribution $\rho(x)$ satisfying the stationary Frobenius–Perron equation, (10.2). To describe a diffusion process produced by this map, we consider the mean square displacement $\langle(\Delta x_n)^2\rangle = \langle(x_n - x_0)^2\rangle$, where the angular brackets denote an average over the equilibrium distribution of initial values, x_0, $\rho(x_0)$.

If the map is such that the mean square displacement grows linearly with the number of iterations or time steps, n, then the diffusion coefficient D is given by

$$
\begin{aligned}
D &= \lim_{n\to\infty}\lim_{L\to\infty}\frac{1}{2n}\left\langle(\Delta x_n)^2\right\rangle \\
&= \lim_{n\to\infty}\lim_{L\to\infty}\frac{1}{2n}\left\langle(x_n - x_{n-1} + x_{n-1} - x_{n-2} + \cdots - x_0)^2\right\rangle \\
&= \lim_{n\to\infty}\lim_{L\to\infty}\frac{1}{2n}\left\langle\left(\sum_{j=0}^{n-1} v_j\right)^2\right\rangle, \qquad (14.1)
\end{aligned}
$$

where we have defined a velocity at each step by $v_j = x_{j+1} - x_j$. If we now evaluate the square of the sum and use the invariance of the equilibrium distribution function, we can write the diffusion coefficient as

$$
D = \frac{1}{2}\left\langle v_0^2\right\rangle + \lim_{n\to\infty}\lim_{L\to\infty}\frac{1}{n}\sum_{j=1}^{n-1}\sum_{j'=0}^{j-1}\langle v_j v_{j'}\rangle . \qquad (14.2)
$$

The terms in the sum in the right-hand side of (14.2) only depend on the time interval $j - j'$ due to the invariance of the equilibrium averages. Then for large enough n we can easily transform this expression into

$$
D = \frac{1}{2}\left\langle v_0^2\right\rangle + \lim_{n\to\infty}\sum_{k=1}^{n-1}\left(1 - \frac{k+1}{n}\right)\langle v_k v_0\rangle , \qquad (14.3)
$$

and we tacitly assume that the limit $L \to \infty$ is understood. If the velocity autocorrelation function $\langle v_k v_0\rangle$ vanishes sufficiently rapidly for large k, we obtain the discrete form of the Green–Kubo formula

$$
\begin{aligned}
D &= \frac{1}{2}\left\langle v_0^2\right\rangle + \sum_{k=1}^{\infty}\langle v_k v_0\rangle \qquad (14.4) \\
&= \sum_{k=0}^{\infty}\langle v_k v_0\rangle - \frac{1}{2}\left\langle v_0^2\right\rangle . \qquad (14.5)
\end{aligned}
$$

This is the Green–Kubo expression for the diffusion coefficient for a particle whose motion is governed by a one-dimensional map. Clearly, similar formulae can be derived for any number of dimensions.

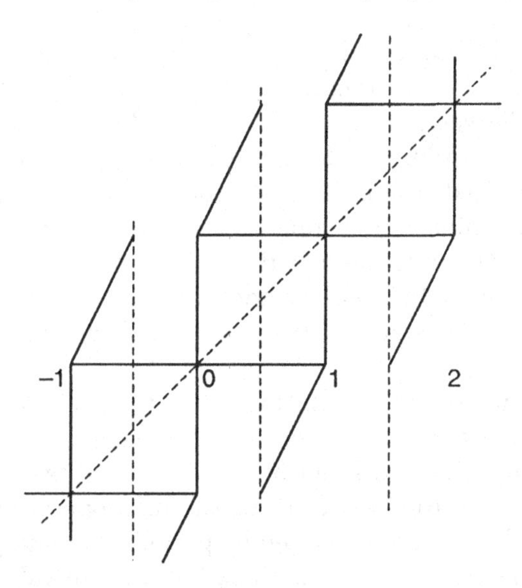

Fig. 14.1 A one-dimensional map that is a deterministic model for a random walk on a line.

14.2 A simple map and a simple fractal

We now apply the Green–Kubo formula to the calculation of the diffusion coefficient for a particle acted upon by a simple map described in Exercise 8.1. This is a one-dimensional map that provides a deterministic realization of a random walk on a line with equal probabilities of moving to the right or to the left at each step. The explicit form of the map is:

$$x_{n+1} = \begin{cases} [x_n] + 1 + 2\tilde{x}_n & \text{for } 0 \leq \tilde{x}_n < \frac{1}{2}, \\ [x_n] - 1 + 2(\tilde{x}_n - \frac{1}{2}) & \text{for } \frac{1}{2} \leq \tilde{x}_n < 1, \end{cases} \tag{14.6}$$

where $[x]$ is the integer part of x, and $\tilde{x}$ is the fractional part of x (see Fig. 14.1). Using the definition of the velocity given above, we find

$$v_n = \begin{cases} 1 + \tilde{x}_n & \text{for } 0 \leq \tilde{x}_n < \frac{1}{2}, \\ -1 + (\tilde{x}_n - 1) & \text{for } \frac{1}{2} \leq \tilde{x}_n < 1, \end{cases} \tag{14.7}$$

Now we make a simplification of this result for the velocity. We are going to set the velocity equal to $\tilde{v}_n = 1$ for $0 \leq x_n < 1/2$, and $\tilde{v}_n = -1$ for $1/2 \leq \tilde{x}_n < 1$. Physically, this is motivated by the fact that the random walk process which we are simulating deterministically by this map just has jumps of ± 1 with equal probability at every step. Thus we might expect that the 'deterministic' corrections to the velocity in (14.7) will not be important when we compute the diffusion coefficient. Mathematically, one can show that this is indeed true by relating the diffusion coefficient to the trace of a Frobenius–Perron operator which arises naturally in this context.† For the case described here, the map is simple enough that one can verify this replacement directly, by computing D using the exact expression for the velocity at each step (see Exercise 14.1).

We now turn to the evaluation of D, using (14.4) and the velocity $\tilde{v}_n$ given above. This evaluation requires us to know the equilibrium distribution function, i.e., the equilibrium solution of the Frobenius–Perron equation for the map operating on the interval $[-L, L]$, and supplemented by periodic boundary conditions applied in an obvious way at each end of the interval. For the map considered here, the equilibrium distribution is uniform on this interval,

$$\rho_{\text{eq}}(x) = \frac{1}{2L}.$$

Further, since we are using periodic boundary conditions, or more generally, since we will eventually take the limit $L \to \infty$, we can easily express D for this map, as

$$D = \lim_{N\to\infty} \sum_{n=0}^{N} \int_0^1 dx \tilde{v}_0(x)\tilde{v}_n(x) - \frac{1}{2}\int_0^1 dx \tilde{v}_0^2(x), \tag{14.8}$$

where we have taken the interval $[0, 1]$ as the starting interval for the particle, since all unit intervals are equivalent when starting the particle at $t = 0$, and we have used the fact that the equilibrium distribution is uniform.

Now we do something very risky, but quite useful nevertheless. We are going to move the summation inside the integral, and define a new function – call it a *jump function* – $J_N(x)$, by

$$J_N(x) = \sum_{n=0}^{N} \tilde{v}_n(x). \tag{14.9}$$

† For details, see the papers by Kadanoff and Tang, and by Cvitanović, Gaspard, and Schreiber listed in *Further reading*.

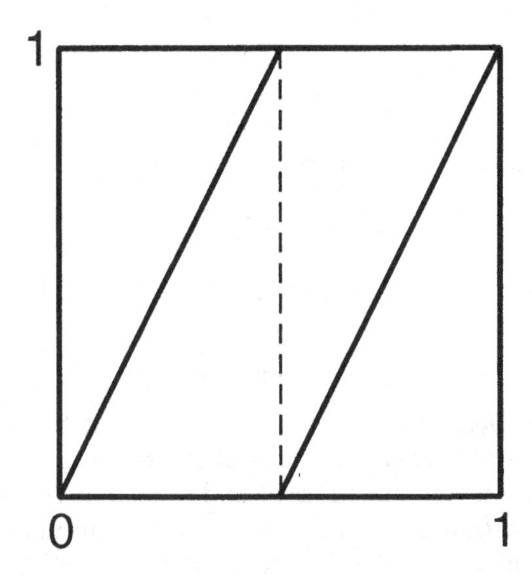

Fig. 14.2 The reduced *modulo* 1 map for the random walk model in Fig. 14.1.

This function is simply the integer value of displacement of the particle from the origin after a total of N steps, having started at the point x. We expect this function to be a very complicated function of x since the final displacement of a point will vary wildly from one initial to the next, as one can check with a simple computer program for this map (see Exercise 14.2). Note also that $J_N(x)$ satisfies a very simple recursion relation

$$J_N(x) = \tilde{v}_0(x) + J_{N-1}\left(\tilde{M}(x)\right), \qquad (14.10)$$

where $\tilde{M}(x)$ is the reduced, or *modulo* 1, map for the map given in (14.6). That is, $\tilde{M}(x)$ maps the unit interval onto itself, and is obtained by taking all the *modulo* 1 values for all of the terms in (14.6). For our map, $\tilde{M}$ is given by (see Fig. 14.2):

$$\tilde{M}(x) = \begin{cases} 2x & \text{for } 0 \le x < \frac{1}{2}, \\ 2x - 1 & \text{for } \frac{1}{2} \le x < 1. \end{cases} \qquad (14.11)$$

One might try to construct a more well-behaved function of x than $J_N(x)$ by expressing J_N as the derivative of a new function

$T_N(x)$:

$$J_N(x) = \frac{d}{dx} T_N(x).$$

Then $T_N(x)$ satisfies the recursion relation

$$T_N(x) = \tilde{t}(x) + \frac{1}{2} T_{N-1}\left(\tilde{M}(x)\right), \tag{14.12}$$

where we have used the fact that the slope of the map $\tilde{M}$ is 2. Here,

$$\tilde{t}(x) = x\tilde{v}(x) + \tilde{c}(x),$$

where $\tilde{c}(x)$ is taken to be constant on each of the sub-intervals where $\tilde{v}$ is constant. We fix the constants $\tilde{c}$ by requiring that $T_N(x)$ be continuous on the unit interval and vanish at the end points – this latter condition is only a convenience. For our map, it is indeed possible to construct such a function. Its limit as N becomes large is a continuous, nowhere differentiable function, called a Takagi function, after the Japanese mathematician who first defined it in 1903. One can immediately see that for our map, the limiting function

$$T(x) = \lim_{N \to \infty} T_N(x)$$

satisfies the recursion relation

$$T(x) = \begin{cases} x + \frac{1}{2} T(2x) & \text{for } 0 \le x < \frac{1}{2}, \\ 1 - x + \frac{1}{2} T(2x - 1) & \text{for } \frac{1}{2} \le x < 1. \end{cases} \tag{14.13}$$

Note that with this construction, $T(x)$ takes the value 0 at the points $x = 0, 1$, and the value $1/2$ at the point $x = 1/2$, regardless of how one approaches $x = 1/2$. The Takagi function can be calculated using a simple iteration scheme, and it is plotted in Fig. 14.3. It is clearly a self-similar fractal and has the properties described above. It is very interesting to note that this fractal underlies the simple random walk on a line. Random walk theory continues to be a rich subject for the exercise of one's imagination!

We can now combine everything above – the Green–Kubo formula, and the Takagi function – to compute the diffusion coefficient for this map. We write

$$D = \int_0^1 dx \tilde{v}_0(x) \frac{d}{dx} T(x) - \frac{1}{2} \int_0^1 dx \tilde{v}_0^2(x). \tag{14.14}$$

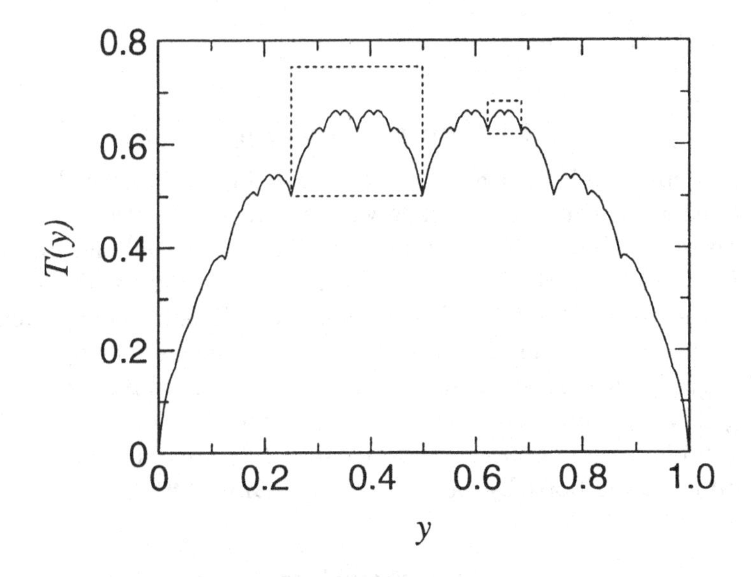

Fig. 14.3. The Takagi function for the random walk map.

If we use the explicit values of the velocity and integrate (notice that the velocity is discontinuous!), we easily find

$$D = 2T(\frac{1}{2}) - T(0) - T(1) - \frac{1}{2} = \frac{1}{2}. \tag{14.15}$$

This is a very complicated way to compute the diffusion coefficient for a simple random walk, but it is very interesting indeed, and can be generalized to describe much more complicated processes where the answers are not so obvious.† The diffusion coefficient is a fractal function of the slope of the map. Takagi functions of the type discussed here arise naturally in the theory of nonequilibrium stationary states for a simple periodic map related to the baker's transformation, as discussed in the paper of Tasaki and Gaspard, listed below. This is part of a more general discussion of the eigenfunctions and eigenvalues of the Frobenius–Perron equation. The eigenvalues of the Frobenius–Perron operator are related in turn to Ruelle–Pollicott resonances and we refer to the literature for

† The Ph.D. thesis of Klages, referred to below, contains a discussion of the close connection of the fractal Takagi function and its generalizations to the fractal forms of diffusion coefficients that appear in simple one-dimensional maps with piecewise linear behavior.

further details.

14.3 Further reading

I am indebted to Rainer Klages and Pierre Gaspard for much of the inspiration for this chapter. The various Takagi functions are discussed in Tasaki and Gaspard [TG95], Klages [Kla96] and in Gaspard's book [Gas98]. This book also contains a discussion of and references to the Ruelle–Pollicott resonances. The idea of replacing the largest eigenvalue of an operator by the trace of the operator, for large times at least, is discussed in the papers of Kadanoff and Tang [KT84], and Cvitanović, Gaspard, and Schreiber [CGS92]. An excellent discussion of one-dimensional dynamical systems can be found in the book by Collet and Eckmann [CE80].

14.4 Exercises

1. Show that the diffusion coefficient calculated by means of Takagi functions remains the same when the exact velocity of the moving particle is used instead of the simplified velocity $v = \pm 1$. You will need to define a new 'Takagi function', but otherwise the calculation is straightforward.

2. Compute and plot the jump function $J_{10}(x)$ for the map discussed here. Do the same using the exact velocity instead of the simplified velocity.

15
Unstable periodic orbits

In the course of our discussions of the baker's map, we noticed that we could easily use its isomorphism with the Bernoulli sequences to locate periodic orbits of the map. As we show below, we can exploit this isomorphism to prove that periodic orbits of the baker's map form a dense set in the unit square. Moreover, we will prove, without much difficulty, that the periodic orbits of the hyperbolic toral automorphisms are also dense in the unit square (or torus). A natural question to ask is: If these periodic orbits are ubiquitous, can they be put to some good use? In this chapter, we outline some simple affirmative answers to this question in the context of nonequilibrium statistical mechanics. In particular, we will see that *periodic orbit expansions* are natural objects when one encounters the need for the trace of a Frobenius–Perron operator, and when one wants to make explicit use of an (ϵ, T)-separated set. Moreover, the periodic orbits of a classical system form a natural starting point for a semi-classical version of quantum chaos theory. We should also mention that there is a new field of study dealing with issues related to the control of chaos, which exploits the presence of periodic orbits to slightly perturb a system from chaotic behavior to a more easily controlled periodic behavior.

15.1 Dense sets of unstable periodic orbits

Here we consider a hyperbolic system. If we have located a periodic orbit of our system, then each point on it has a set of stable and unstable directions. Therefore any nearby point (with the exception of a set of Lebesgue measure zero), will move away from the periodic orbit points in the course of time. Thus all periodic orbits of a hyperbolic system are *unstable* in this sense.

The proof that periodic points are dense in the unit square for the baker's map is very simple. Consider some point, (x, y), in the unit square, and express its value in binary notation as

$$(x, y) = \cdots b_{-N} b_{-N+1} b_{-N+2} \cdots b_{-1} . a_0 a_1 \cdots a_N \cdots . \tag{15.1}$$

Here, the a_i and b_{-j} are each either 0 or 1 in the binary decimal notation. As indicated, we have specified in (15.1), $2N+1$ digits in the binary representation of the point (x, y), while the representation may proceed to infinity in both directions. However, the block of $2N+1$ digits specifies the x and y values of the point to a precision of order 2^{-N}. If N is large, this locates the point very precisely. Now, using this block of $2N+1$ digits, we can construct a new point, (x', y'), say, which for large enough N is arbitrarily close to (x, y), and which is also a periodic point of the baker's map, with period $2N+1$ or less. To construct (x', y'), we take the block of digits from b_{-N} to a_N, and repeat them to the left and right, over and over, *ad infinitum*. In this way,

$$(x', y') = \cdots a_{N-1} a_N b_{-N} b_{-N+1} \cdots b_{-1} . a_0 a_1 \cdots a_N b_{-N} b_{-N+1} \cdots . \tag{15.2}$$

Recalling that the baker's transformation is a shift of the entire sequence one unit to the left, we see that (x', y') is a periodic point of period at most $2N+1$, and that it is arbitrarily close to (x, y) for large enough N. We say that the period is at most $2N+1$, because the block of $2N+1$ digits may itself be periodic with a shorter period.

The argument that periodic points for hyperbolic toral automorphisms are dense on the unit torus is not much more difficult. Recall that such an automorphism is defined, in two dimensions, by the matrix T of the form

$$T = \begin{bmatrix} a & b \\ c & d \end{bmatrix}, \tag{15.3}$$

where a, b, c, d are integers, $\det T = 1$, and the eigenvalues are not equal to unity. Using the fact that any power of the matrix T is also a hyperbolic toral automorphism, we can argue that any solution (x, y) of the equation

$$T^n \cdot \begin{bmatrix} x \\ y \end{bmatrix} = \begin{bmatrix} x \\ y \end{bmatrix} \tag{15.4}$$

must have rational values for both x and y, since only integers occur in the linear, algebraic equations for x and y. Thus we know that periodic points are composed of rational components. A similar but somewhat more lengthy argument (see Exercise 15.1) shows that *any* point with rational x, y is a periodic point of T.

The results just obtained are simple examples of a broader result that, under very general conditions, periodic points are dense in the phase-space for a hyperbolic system.

15.2 The topological zeta-function

It is time, now, to introduce some terms. We say that a *periodic orbit* of period n consists of the *periodic points* $\Gamma_0, \Gamma_1, \Gamma_2, \ldots, \Gamma_{n-1}$, each point of which is transformed to the next successive point under the transformation being considered. A periodic orbit may be *prime* or it may be *reducible*. A reducible orbit of period n is composed of prime orbits of shorter periods, very much as an integer can be expressed as a product of prime factors. For example, the period 4 binary sequence 1010, repeated indefinitely, is really composed of a period 2 sequence 10, while the period 4 sequence 1000 is prime. Thus if a prime orbit has period n_{p}, then there are periodic orbits, of period rn_{p}, with r a positive integer, that can be generated from it by taking r repetitions.

We call $P_n(M)$ the number of periodic *points* of period n of the map M. There are some simple formulae for $P_n(M)$ for simple maps. For example, we argued in the problems for the baker's map, that $P_n(B) = 2^n - 1$. For a two-dimensional hyperbolic toral automorphism T, one can show that $P_n(T) = \exp(n\lambda_+) + \exp(n\lambda_-) - 2 = 2[\cosh(n\lambda_+) - 1]$, where $\lambda_\pm$ are the positive and negative Lyapunov exponents, and $\lambda_- = -\lambda_+$.

Since there seems to be a nice analogy between periodic orbits and prime numbers, we might try to introduce some techniques from number theory in the study of periodic orbits. In particular, we can introduce a *topological zeta-function*, $\zeta_{\mathrm{top}}(z)$, by

$$\zeta_{\mathrm{top}}(z) = \exp\left[\sum_{n=1}^{\infty} P_n(M)\frac{z^n}{n}\right]. \tag{15.5}$$

This zeta-function can be expressed in terms of prime orbits by noting that if an orbit of period n is not a prime orbit, then $n = rT_{\mathrm{p}}$ where r is the number of repetitions of the appropriate prime orbit of period T_{p}. Then the number of distinct points of the prime orbit is $T_{\rm p}$. Consequently, we can reorganize the sum appearing in the exponent in (15.5) as

$$\sum_{1}^{\infty} P_n(M)\frac{z^n}{n} = \sum_{p}\sum_{r=1}^{\infty} T_{\rm p}\frac{z^{rT_{\rm p}}}{rT_{\rm p}} = -\sum_{\rm p}\ln(1-z^{T_{\rm p}}), \qquad (15.6)$$

where the sum is over all distinct prime periodic orbits. Using this result, we can give a general expression for the topological zeta-function

$$\zeta_{\rm top}(z) = \prod_{\rm p}\left[1-z^{T_{\rm p}}\right]^{-1}, \qquad (15.7)$$

where the product is over all distinct prime periodic orbits. For example, the periodic repetition of the sequence 0011 is the same sequence as that generated by 1001, 1100, and 0110. However, this sequence differs from another prime orbit of period 4 generated by $0001 = 1000 = 0100 = 0010$.† The product over prime orbits is indeed reminiscent of the expression for the Riemann zeta-function, familiar from number theory. The Riemann zeta-function, $\zeta_{\rm R}(s)$ is defined, for $s > 1$, by the relation

$$\zeta_{\rm R}(s) = \sum_{n=1}^{\infty}\frac{1}{n^s},$$

and it is easily transformed to an infinite product over prime numbers, p, as

$$\zeta_R(s) = \prod_{p}(1-\frac{1}{p^s})^{-1} = \prod_{p}(1-z^{\ln p})^{-1}, \qquad (15.8)$$

where $z = \exp(-s)$. This identity suggests that the Riemann zeta-function is the topological zeta-function for a 'dynamical system' whose prime periods are the logarithms of the prime numbers.‡

Finally, we can use the known expressions for $P_n(M)$ to give explicit expressions for the topological zeta-functions for the baker's

† Often one denotes a particular prime orbit with a line over the repeating sequence as in $\overline{0011} = \overline{1001}$, etc. Note also that there are always $T_{\rm p}$ distinct points in an orbit of period $rT_{\rm p}$, no matter the value of r.

‡ I am indebted to Dr Carl Dettmann for this amusing observation.

map and the Arnold cat map. For the baker's map we have

$$\begin{aligned}\zeta_{\text{top}}(z) &= \exp\left[\sum_{n=1}^{\infty}(2^n-1)\frac{z^n}{n}\right]\\ &= \frac{1-z}{1-2z},\end{aligned} \tag{15.9}$$

and for a hyperbolic toral automorphism,

$$\begin{aligned}\zeta_{\text{top}}(z) &= \exp\left[\sum_{n=1}^{\infty}\left[\exp(n\lambda_+)+\exp(n\lambda_-)-2\right]\frac{z^n}{n}\right]\\ &= \frac{(1-z)^2}{[1-z\exp(\lambda_+)][1-z\exp(\lambda_-)]}.\end{aligned} \tag{15.10}$$

On comparing the formal expression for the zeta-function, (15.7), with the explicit results given by (15.9) and (15.10), we note that infinite products may have unexpected forms when evaluated. Notice also that these two zeta-functions are meromorphic functions in the complex plane with simple poles on the positive real axis. The reader should notice that the zeta-function has a pole at $\ln z = -h_{\text{top}}$.† For the baker's map, the topological entropy is $h_{\text{top}} = \ln 2$ and for the hyperbolic toral automorphism, $h_{\text{top}} = \lambda_+$.

15.3 Periodic orbits and diffusion

One of the reasons for the wide-spread interest in periodic orbits of a dynamical system is that, under favorable circumstances, a number of the orbits can be located, either analytically or using computer methods, and their properties can be determined. As we shall see below, one important property of a periodic orbit is the set of Lyapunov exponents associated with it. Until now we have had no occasion to define the Lyapunov exponent for a periodic orbit, but there is a natural definition which is quite useful. We will present this definition below, but first some motivation. We show below that it is possible to compute escape rates and diffusion coefficients using expansions in terms of periodic orbits. While we will confine ourselves to the case of diffusion, it is not hard to show

† We will see in a subsequent section that the topological entropy can be expressed as the rate of growth of the number of periodic orbits of period n as $n \to \infty$.

how the arguments can be extended to other transport processes as well.

15.4 Escape-rates and periodic orbits

Let us suppose that we have a particle located at a point, Γ, in a region of phase-space, denoted by $\mathcal{V}$, which has absorbing boundaries. As usual in escape-rate methods, if the particle crosses the absorbing boundaries, it is removed from the system. Suppose further that at each time step the particle is acted upon by a map $M(\Gamma)$ that moves the point to a new point in phase-space. At the initial time, we suppose that the probability distribution for the particle is uniform over $\mathcal{V}$, and we ask for the probability that the particle is still in $\mathcal{V}$ after n steps. This probability will decrease exponentially in time, if M is hyperbolic, and the probability we seek should decrease in time as $\exp(-n\gamma)$ for large n, and γ is the escape rate.

A heuristic expression for the escape rate is obtained if we express the probability that the point is still in $\mathcal{V}$ after n steps as

$$e^{-n\gamma} = \frac{\int_{\mathcal{V}} d\Gamma \int_{\mathcal{V}} d\Gamma' \delta(\Gamma - M^n(\Gamma'))}{\int_{\mathcal{V}} d\Gamma'}, \tag{15.11}$$

where we have included a normalization factor so that the right-hand side of (15.11) is unity at $n = 0$. Let us now introduce an integral operator, the *Frobenius–Perron operator*, $\mathcal{L}(\Gamma_2|\Gamma_1) = \delta(\Gamma_2 - M(\Gamma_1))$. Powers of $\mathcal{L}$ are defined by the usual iteration equation

$$\begin{aligned} \mathcal{L}^2(\Gamma_3|\Gamma_1) &= \int_{\mathcal{V}} d\Gamma_2 \delta(\Gamma_3 - M(\Gamma_2))\delta(\Gamma_2 - M(\Gamma_1)) \\ &= \delta(\Gamma_3 - M^2(\Gamma_1)). \end{aligned} \tag{15.12}$$

Consequently, we write

$$e^{-n\gamma} = \frac{\int_{\mathcal{V}} d\Gamma \int_{\mathcal{V}} d\Gamma' \mathcal{L}^n(\Gamma|\Gamma')}{\int_{\mathcal{V}} d\Gamma'}. \tag{15.13}$$

Now let us continue to think heuristically, and assume that the operator $\mathcal{L}$ has a complete set of eigenfunctions, with corresponding eigenvalues ω_i. An elementary calculation based on the assumed eigenexpansion of $\mathcal{L}$ shows that the escape rate $\gamma = \ln \omega_{\max}$, where $\omega_{\max}$ is the largest eigenvalue of $\mathcal{L}$. We are not actually going to compute this largest eigenvalue, but instead we use the fact the same result for the escape rate would be obtained if we were to use the *trace* of $\mathcal{L}^n$ and replace (15.11) by

$$e^{-n\gamma} = \int_{\mathcal{V}} d\Gamma \delta(\Gamma - M^n(\Gamma)). \tag{15.14}$$

This formula was first derived by Kadanoff and Tang, and its utility is based upon the fact that the escape rate is now determined by the properties of periodic orbits of period n for large n, where the particle remains in the region $\mathcal{V}$ over this time, since the delta function in (15.14) picks out only these orbits.

The right-hand side of (15.14) can be expressed in terms of the Jacobian of the map M^n at the periodic points by expanding the argument of the delta function about its zero points. Then we find that

$$e^{-n\gamma} = \sum_{\Gamma_{i,n}} \frac{1}{|\det[\mathbf{1} - \mathbf{J}^{(n)}(\Gamma_{i,n})]|}, \tag{15.15}$$

where the sum is over all points, $\Gamma_{i,n}$, of period n for those orbits which remain in the indicated region, and the Jacobian of the nth iterate of the map M is

$$\mathbf{J}^{(n)}(\Gamma_{i,n}) = \prod_{j=0}^{n-1} \mathbf{J}(M^j(\Gamma_{i,n})), \tag{15.16}$$

where $\mathbf{J}(\Gamma)$ is the Jacobian of the map M at Γ. The determinant appearing in the denominator on the right-hand side of (15.15) can be expressed in terms of eigenvalues, denoted by $\Lambda_\alpha^{(n)}$, of the Jacobian $\mathbf{J}^{(n)}$, as

$$\left|\det\left[\mathbf{1} - \mathbf{J}^{(n)}(\Gamma_{i,n})\right]\right| = \prod_\alpha \left|(1 - \Lambda_\alpha^{(n)})\right|. \tag{15.17}$$

These eigenvalues can be obtained from computer studies by finding the periodic points and following them through their orbits to determine the required Jacobians. It is not easy, especially if n is large, but in many cases it can be done. The eigenvalues provide a natural definition of the Lyapunov exponents, $\lambda_\alpha^{(n)}$, for a point of an orbit of period n as

$$\lambda_\alpha^{(n)} = \frac{1}{n} \ln \left|\Lambda_\alpha^{(n)}\right|. \tag{15.18}$$

This is the definition to which we referred at the beginning of this section.

In a way that is very much like the case for the Lyapunov exponents for a point that is not on a periodic orbit, the Lyapunov exponents for periodic points can be grouped in a set of positive, expanding, exponents and a set of negative, contracting, exponents. This means that the escape rate can be expressed in terms of the positive Lyapunov exponents alone, since for large n, the negative Lyapunov exponents will be associated with values of $\Lambda_\alpha^{(n)}$ which are small compared to unity, and can be neglected in the above determinant. Similarly, the $\Lambda_\alpha^{(n)}$ associated with positive Lyapunov exponents will be very large compared to unity, so that for large n,

$$e^{-n\gamma} = \sum_{\Gamma_{i,n}} \prod{}' \frac{1}{|\Lambda_\alpha(\Gamma_{i,n})|}, \tag{15.19}$$

where the prime on the product denotes the product only over the expanding directions, and we explicitly indicate the possible dependence of the eigenvalues on the periodic points, $\Gamma_{i,n}$. This result is consistent with our statements in Chapter 13 that only the expanding directions are important in the construction of Gibbs measures for a dynamical system. We will return to this point in the next section.

Our final result for the escape rate, (15.19), depends on a number of things, in addition to the positive Lyapunov exponents for the periodic points. It depends on finding the orbits of period n that remain within $\mathcal{V}$, which, in turn, depend upon the exact structure of the region $\mathcal{V}$. However, once the escape rate is determined, a diffusion coefficient can be computed in the way indicated in Chapter 12.

We conclude this section by deriving a very useful identity for the spectrum of the Frobenius–Perron operator, as an expansion in terms of the properties of the periodic orbits. We are interested in the eigenvalues of the operator $\mathcal{L}(\Gamma_2|\Gamma_1)$. Of course, these eigenvalues will be the inverses of the roots, z, of the Fredholm determinant $\det(1 - z\mathcal{L})$. Now let's express this determinant in terms of the traces of powers of $\mathcal{L}$ as

$$\begin{aligned}
\det(1 - z\mathcal{L}) &= \exp[\operatorname{tr}\ln(1 - z\mathcal{L})] \\
&= \exp\left[-\sum_{n=1}^{\infty} z^n \frac{\operatorname{tr}\mathcal{L}^n}{n}\right] \\
&= 1 - z\operatorname{tr}\mathcal{L} - \frac{z^2}{2}\left(\operatorname{tr}\mathcal{L}^2 - (\operatorname{tr}\mathcal{L})^2\right) + \cdots .
\end{aligned} \tag{15.20}$$

Here, we have expanded the logarithm in the exponent and then expanded the exponent in a power series in z. As before, we can now expand the trace of powers of the Frobenius–Perron equation in terms of the properties of the periodic orbits, so that

$$\operatorname{tr}\mathcal{L}^n = \sum_{\Gamma_{i,n}} \frac{1}{|\det[\mathbf{1} - \mathbf{J}^{(n)}(\Gamma_{i,n})]|}. \tag{15.21}$$

Here, again, when one evaluates the determinant, one need only keep the terms which depend upon the product of the expanding eigenvalues, which, in turn, depend on the periodic points, $\Gamma_{i,n}$. The Fredholm determinant method is useful for developing techniques to speed the convergence of periodic orbit expansions for various dynamical quantities. In Exercise 15.2, we suggest a method to obtain a simple series of approximations to the spectrum of the Frobenius–Perron operator, using (15.21) and (15.21).

15.5 The generating function method for diffusion

There is another way to use periodic orbit expansions to obtain an expression for the diffusion coefficient, that is similar to the method given above but does not rely on the construction of a region with absorbing boundaries, and the many complications that result from doing so. The method described below is particularly well-suited to systems with a periodic configuration space, such as a Lorentz gas with a periodic placement of scatterers, as illustrated in Fig. 15.1, or a chain of one-dimensional maps such as the chain studied in Chapter 14. We suppose for simplicity that we are dealing with a discrete-time map, M, rather than a continuous-time flow.

We recall that the diffusion coefficient is related to the mean

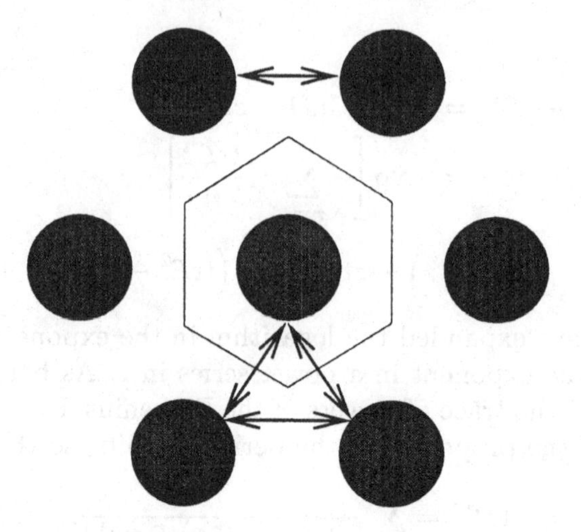

Fig. 15.1 Some simple periodic orbits with period 2 and period 3 along with the unit cell for the triangular Lorentz gas.

square displacement of the moving particle in the x-direction, say,

$$D = \lim_{n\to\infty} \frac{1}{2n} \left\langle (x(n) - x(0))^2 \right\rangle, \tag{15.22}$$

where the average denoted by the angular brackets is over some equilibrium distribution for the initial points, $x(0)$. Here, $x(n)$ is the displacement after the nth time step of the point that started at $x(0)$. We can obtain the mean square displacement by considering a more general function, $Q(\sigma)$, defined by

$$Q(\sigma) = \lim_{n\to\infty} \frac{1}{n} \ln \langle \exp\{\sigma[x(n) - x(0)]\}\rangle, \tag{15.23}$$

where the average of $\exp\{\sigma[x(n) - x(0)]\}$ may be considered as a generating function for diffusion.

Now let us suppose that the entire phase-space, denoted by X, can be decomposed into a periodic tiling of unit cells and that the initial points are distributed uniformly in one particular unit cell which we denote by C. Then the generating function may be written similarly to (15.11) as

$$\begin{aligned} &\langle \exp\{\sigma[x(n) - x(0)]\}\rangle \\ &= \frac{1}{\int_C d\Gamma} \int_C d\Gamma \int_X d\Gamma' \exp\{\sigma[x(\Gamma') - x(\Gamma)]\} \times \\ &\qquad \delta(\Gamma' - M^n(\Gamma)), \end{aligned} \tag{15.24}$$

where $x(\Gamma)$ is the x-coordinate of the point at phase-space point Γ. Since the full phase-space consists of a periodic arrangement (in configuration space) of unit cells, we can keep track of the particle by expressing its position $\mathbf{r}$ after n iterations of the map by writing

$$\mathbf{r}(\Gamma') = \nu_i(\Gamma') + \tilde{\mathbf{r}}(\Gamma'), \tag{15.25}$$

where $\nu_i(\Gamma')$ is the location of the particular unit cell, i, in which a point at phase Γ' is located, and $\tilde{\mathbf{r}}(\Gamma')$ is the location of the point at Γ' with respect to some reference position within a unit cell. We suppose that the reference point for the starting cell, C, is the origin of our coordinate system. Notice that $\tilde{\mathbf{r}}(\Gamma')$ ranges only within a unit cell.

With this notation, we can express the Γ' integral over the full phase-space in (15.24) as a sum over all the possible unit cells, i, together with an integral over $\tilde{\Gamma}'$ that ranges only over a unit cell. This enables us to write (15.24) as

$$\langle \exp\{\sigma[x(n) - x(0)]\}\rangle = \frac{\int_C d\Gamma \int_C d\tilde{\Gamma}' \mathcal{L}_\sigma^{(n)}(\tilde{\Gamma}'|\Gamma)}{\int_C d\Gamma}. \tag{15.26}$$

We have now defined a generalized Frobenius–Perron operator $\mathcal{L}_\sigma^{(n)}(\tilde{\Gamma}'|\Gamma)$ by

$$\begin{aligned} \mathcal{L}_\sigma^{(n)}(\tilde{\Gamma}'|\Gamma) &= \sum_i \exp\left\{\sigma\left[\nu_x(\Gamma') + x(\tilde{\Gamma}') - x(\Gamma)\right]\right\} \times \\ &\quad \delta(\Gamma' - M^n(\Gamma)) \\ &= \exp\left\{\sigma\left[\nu_x(\Gamma, n) + (x(\tilde{\Gamma}') - x(\Gamma))\right]\right\} \times \\ &\quad \delta(\tilde{\Gamma}' - \tilde{M}^n(\Gamma)). \end{aligned} \tag{15.27}$$

Here we have summed over all the unit cells in the tiling of the phase-space, and set $\nu(\Gamma, n)$ as the location of the unit cell in which a point starting at Γ will find itself after n iterations of the map. We have also defined a reduced map $\tilde{M}$ which follows points within a unit cell but pays no attention to the particular unit cell the particle is in after any action of the full map, M. This corresponds to the *modulo* 1 map introduced in Sec. 14.2.

As in the calculation of the escape rate in the previous subsection, and for exactly the same reasons, we replace the calculation of the numerator on the right-hand side of (15.26) by a trace, so that for large n we may write

$$\begin{aligned}\langle \exp\{\sigma[x(n) - x(0)]\}\rangle &= \exp[nQ(\sigma)] \\ &= \int_C d\Gamma \exp[\sigma\nu_x(\Gamma, n)]\,\delta(\Gamma - \tilde{M}^n(\Gamma)).\end{aligned} \tag{15.28}$$

The delta function now picks out the points in the unit cell which are period n points of the unit cell map $\tilde{M}$, but we still have to keep track of the unit cell, $\nu(\Gamma, n)$ for the point Γ after n iterations of the full map, M. By evaluating the delta function in terms of the properties of the map at the periodic points $\Gamma_{j,n}$, we find that for large n

$$\begin{aligned}\exp[nQ(\sigma)] &= \sum_{\Gamma_{i,n}} \frac{\exp[\sigma\nu_x(\Gamma_{i,n})]}{|\det[\mathbf{1} - \mathbf{J}^{(n)}(\Gamma_{i,n})]|} \\ &= \sum_{\Gamma_{i,n}} \frac{\exp[\sigma\nu_x(\Gamma_{i,n})]}{\prod_\alpha |(1 - \Lambda_\alpha^{(n)})|}.\end{aligned} \tag{15.29}$$

Here we have introduced, as before, the Jacobian of reduced map, $\tilde{M}^n$, and its eigenvalues, $\Lambda_\alpha^{(n)}$, labeled by the index α.

Equation 15.29 is the periodic orbit expansion of the generating function $Q(\sigma)$, from which the diffusion coefficient D can be obtained by taking the second derivative of Q with respect to σ and setting $\sigma = 0$. We here assume that there is no average drift of the particle, so that all odd derivatives of Q with respect to σ vanish when $\sigma = 0$. The right-hand side of (15.29) can be expressed in terms of a zeta-function similar to, but somewhat more complicated than, the topological zeta-function introduced earlier in this chapter. The new zeta-function is called a *Ruelle zeta-function*, and its properties are the subject of a large literature, to which we refer the reader for details. We mention here that the methods used in this section can be applied to computing the averages of many dynamical quantities, including Lyapunov exponents.

15.6 Periodic orbits and Gibbs measures

We conclude this brief introduction to periodic orbit expansion methods with a discussion of the application of periodic orbit points to the construction of Gibbs measures.

In order to construct a Gibbs measure on a phase-space we need a set of (ϵ, T)-separated points, for some small ϵ and large time T, and the expansion factors corresponding to those points. These expansion factors are then used to specify a weight for the small phase-space region containing one of these points and to compute the dynamical partition function, $Z(\beta)$. Therefore, we can consider all of the periodic points of period N for large N and regard these as the separated set, although we would need to study the orbits carefully in order to specify the appropriate values of ϵ and T precisely. Since the limits of $\epsilon \to 0$ and $T \to \infty$ are taken at the end, we only have to choose the period N large enough and take the limit $N \to \infty$ at the end.

We can associate a Gibbs measure, $w_{i,N}$, with each periodic point, $\Gamma_{i,N}$, as

$$w_{i,N} = \frac{1}{Z_N(\beta)} \exp\left[-\beta N \sum_{\lambda_j>0} \lambda_j(\Gamma_{i,N})\right], \tag{15.30}$$

where the Lyapunov exponents for the periodic orbits, $\lambda_j(\Gamma_{i,N})$ are defined in (15.18), and the dynamical partition function is given by

$$Z_N(\beta) = \sum_{\Gamma_{i,N}} \exp\left[-\beta N \sum_{\lambda_j>0} \lambda_j(\Gamma_{i,N})\right]. \tag{15.31}$$

Given the dynamical partition function and the weights above, one can calculate, to as good an approximation as one wants, the dynamical properties of the system using the general methods of the thermodynamic formalism. For example, using the fact that there are $2^N - 1$ points of period N for the baker's map, and that they all have the same Lyapunov exponent, $\ln 2$, we can easily, for this map, obtain the dynamical partition function,

$$Z_N(\beta) = (2^N - 1)\exp\left[-\beta N \ln 2\right], \tag{15.32}$$

and the topological pressure

$$P(\beta) = \lim_{N\to\infty} \frac{1}{N} \ln Z_N(\beta) = (1-\beta)\ln 2. \tag{15.33}$$

Of course, this result agrees with the previous expression, (13.74), for the topological pressure.

The periodic orbit expansion method has been widely used to compute dynamical properties of many interesting systems such as periodic Lorentz gases, and of systems of particles subjected to external forces and a Gaussian thermostat which maintains a constant kinetic or total energy. A number of useful references are listed below.

15.7 Further reading

A special issue of the journal *Chaos* Vol. 2, No. 1, (1993), edited by P. Cvitanović, has been devoted to periodic orbit theory. Cvitanović [C+97] has also written an on-line book, *Classical and Quantum Chaos: A Cyclist Treatise*, which can be accessed at http://www.nbi.dk/ChaosBook/. A good mathematical description of the use of periodic orbits to construct SRB and Gibbs measures can be found in the paper of Parry [Par88]

Papers on periodic orbit expansions include those by Artuso *et al.* [AAC90a, AAC90b] and by Cvitanović and Eckhardt [CE91]. The applications of these methods to periodic Lorentz gases can be found in Cvitanović *et al.* [CGS92], Vance [Van92], Grebogi, Ott, and Yorke [GOY88], Rondoni and Morriss [RM97], and Morriss *et al.* [MDR97]. A recent paper by Rondoni and Cohen discusses applications of periodic orbit theory to derive Onsager relations, among other things [RC98]. Additional material on periodic orbit expansions for quantum systems can be found in the book by Gutzwiller [Gut90] and in the paper by Gaspard [Gas95] in the book edited by Casati and Chirikov [CC95].

15.8 Exercises

1. Show that any point on the unit torus with rational coordinates (x, y) is a periodic point of a hyperbolic toral automorphism. You may consider the Arnold cat map as a typical example.

2. Use (15.21) and (15.22) to obtain linear and quadratic approximations to the spectrum of the Frobenius–Perron operator for the baker's map and for the Arnold cat map, on the unit torus.

3. Use periodic orbit theory to calculate the topological pressure for the Arnold cat map.

4. Use periodic orbit theory to compute the diffusion coefficient for a particle subjected to the one-dimensional map of (14.6).

16
Lorentz lattice gases

In this chapter, we will discuss briefly some simple models of fluid systems that are designed to exhibit many of the nonequilibrium properties of a real fluid, and to be very suitable for precise computer studies of fluid flows since only binary arithmetic is used to simulate these models. The models were devised by Frisch, Hasslacher, and Pomeau, among others, and are generally called *cellular automata lattice gases.* The corresponding one-dimensional Lorentz gas, studied in great detail by Ernst and co-workers, may be viewed as a 'modern-day' Kac ring model. The interest of these models for us consists in the fact that it is rather straightforward to compute both the transport as well as the chaotic properties of these systems, and the thermodynamic formalism is especially useful here. After introducing the general class of cellular automata lattice gases (CALGs) we will turn our attention to the special case of the one-dimensional Lorentz lattice gas (LLG) to outline how its dynamical quantities can be calculated.

16.1 Cellular automata lattice gases

Consider a two-dimensional hexagonal or square lattice with bonds connecting the nearest-neighbor lattice sites. A CALG is constructed by (i) putting indistinguishable particles on this lattice with velocities that are aligned along the bond directions, (ii) considering that the time is discretized, and (iii) stating that in one time step a particle goes from one site to the next in the direction of its velocity. The number of possible velocities for any particle is then equal to the coordination number, b, of the lattice, although models with rest particles (zero velocity), or with other velocities, are often considered. To complete the specification of the model one has to provide the collision rules governing what happens to the velocities of the particles when two or more of them arrive at the same site from different directions. The collision rules are either deterministic, where the outcomes of such collisions are determined entirely by the incoming velocities, or stochastic, where several outcomes are possible, with a given probability for the occurrence of each possibility. Common to both types of models is an exclusion principle whereby it is impossible for two particles with the same velocity to occupy the same site simultaneously, thus the total number of particles allowed at a site is the coordination number, b.† In Fig. 16.1, we illustrate the stochastic collision rules for particles on a hexagonal lattice in two dimensions, the Frisch–Hasslacher–Pomeau (FHP) model.

The CALG models have proved to be very useful for simulating hydrodynamic flows on a computer and for studying nonequilibrium processes analytically. The most accurate studies of the long-time decays of the equilibrium time–correlation functions appearing in the Green–Kubo formulae have been carried out for these models. Moreover, recently CALGs have provided useful models for studying pattern formation and other hydrodynamical instabilities in fluids.

It is worth pointing out here that while it is very clear how to apply the thermodynamic formalism for chaotic properties to CALGs with probabilistic collision rules, it is not so clear how to do so if the rules are deterministic. This problem is connected to the lack of an underlying differential structure for deterministic CALGs. Probabilistic CALGs can be mapped onto a deterministic dynamical system with a differential structure in very much the same way that a random coin toss system can be mapped onto a baker's map.

16.2 Chaotic behavior of Lorentz lattice gases

In order to show how the methods developed here can be applied to CALGs, we consider the very simplest of these models, the one-dimensional Lorentz lattice gas. We imagine a finite, one-dimensional lattice of points at integer values with either open or

† This exclusion principle simplifies both the computer simulations and the analysis because it prevents a very large number of particles from being at any lattice site, and makes the models stable.

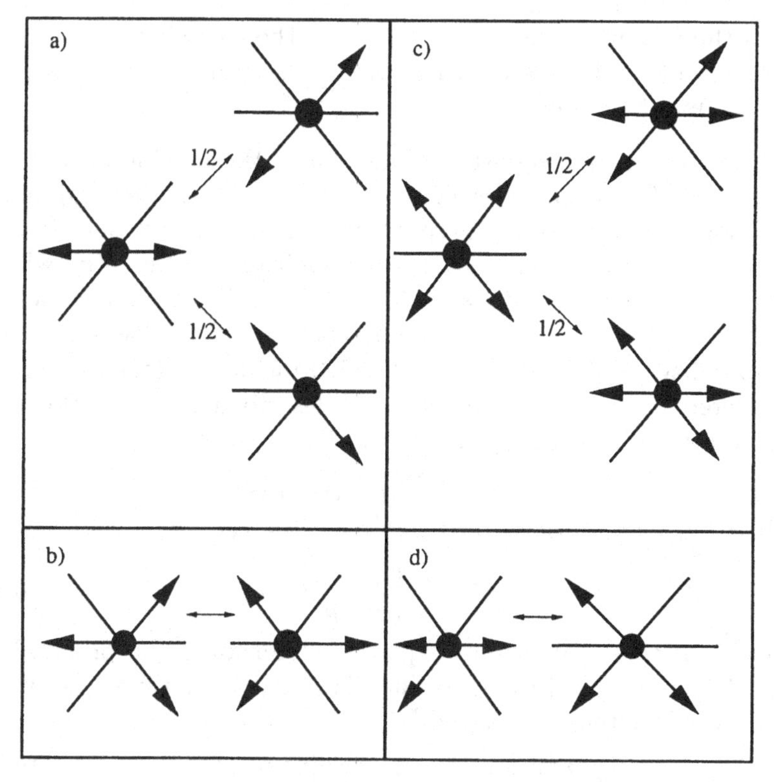

Fig. 16.1 Stochastic collision rules for the FHP model: (a) head-on collision with two equally probable outcomes; (b) triple collision; (c) dual of head-on collision under particle–hole exchange; (d) head-on collision with spectator.

periodic boundaries. One particle moves on this lattice with velocity $\mathbf{c} = \pm 1$. Fixed scatterers are placed at various sites on the lattice, with average density ρ, and the moving particle can collide with them. The collision rules are:

1. A particle with velocity vector $\mathbf{c}$ collides with a scatterer when it arrives at a site on which a scatterer is placed. In such a collision there is a probability p that the particle will maintain velocity $\mathbf{c}$ and continue in the same direction at the next time step, and probability $q = 1 - p$ that the encounter with the scatterer will reverse the direction of the velocity of the particle.

2. If there is no scatterer at a site when the particle arrives there, then it will continue in the same direction, with the same velocity, at the next time step.

The microscopic equation of motion of the moving particle in a fixed configuration of scatterers is described by the Chapman–Kolmogorov equations. To construct them, we first specify the configuration of scatterers by a set of Boolean variables $\hat{\rho}(\mathbf{r})$, where $\mathbf{r}$ is a point on the lattice, such that $\hat{\rho}(\mathbf{r}) = 1$ if there is a scatterer at $\mathbf{r}$ and $\hat{\rho}(\mathbf{r}) = 0$ if there is no scatterer at $\mathbf{r}$. The scattering probability, $\hat{W}_{ij}(\mathbf{r})$, for a particle with incident velocity $\mathbf{c}_j$ to be scattered to velocity $\mathbf{c}_i$ at site $\mathbf{r}$ is described by these Boolean variables, and the scattering probabilities, p, q as

$$\hat{W}_{ij}(\mathbf{r}) = \hat{\rho}(\mathbf{r}) W_{ij} + [1 - \hat{\rho}(\mathbf{r})]\delta_{ij}, \tag{16.1}$$

where W_{ij} is a Markov transition matrix given by

$$W_{ij} = \begin{bmatrix} p & q \\ q & p \end{bmatrix}. \tag{16.2}$$

The Chapman–Kolmogorov equation determines the probability $\pi(x,t)$, with $x \equiv \{\mathbf{r}, \mathbf{c}_i\}$, to find the particle at time t at site $\mathbf{r}$ with incident velocity $\mathbf{c}_i$, and is

$$\pi(x, t+1) = \sum_y w(x|y)\pi(y,t). \tag{16.3}$$

Here, $w(x|y)$ is the transition matrix giving the probability to go from a state $y \equiv \{\mathbf{r}', \mathbf{c}_j\}$ to state $x \equiv \{\mathbf{r}, \mathbf{c}_j\}$, and is given by

$$w(x|y) = \delta(\mathbf{r} - \mathbf{c}_i, \mathbf{r}')\hat{W}_{ij}(\mathbf{r}'). \tag{16.4}$$

Here the delta function is a Kronecker delta.

Although we will not pursue this point in much detail, (16.3) forms the basis of a kinetic theory analysis of the diffusive motion of the moving particle on the lattice. This equation corresponds to a Liouville equation for the moving particle. To obtain a macroscopic equation, we need to average (16.3) over all possible configurations of scatterers consistent with a fixed average density $\rho = \langle\hat{\rho}(\mathbf{r})\rangle$, where the angular brackets denote an average over all possible configurations of scatterers. This averaging can be carried out either by considering all possible arrangements of N scatterers on L sites, with $N/L = \rho$, or by allowing the number of scatterers to fluctuate and simply specify that among all members of the ensemble, the probability of finding a scatterer at any given site is ρ. In any case, the average of (16.4) corresponds to the first BBGKY hierarchy equation, since the average of the right-hand side leads to a pair correlation function for the moving particle and a scatterer. That is, if we write $f(x,t) = \langle \pi(x,t) \rangle$, we obtain

$$\begin{aligned} f(x,t+1) &= \sum_y \langle w(x|y)\pi(y,t)\rangle \\ &= \sum_y \langle w(x|y)\rangle\, f(y,t) \\ &+ \sum_y \langle w(x|y)\pi(y,t)\rangle - \langle w(x|y)\rangle\, f(y,t), \end{aligned} \tag{16.5}$$

where

$$\langle w(x|y)\rangle = \delta(\mathbf{r}-\mathbf{c}_i,\mathbf{r}')[\rho W_{ij} + (1-\rho)\delta_{ij}]. \tag{16.6}$$

We have written the second line of (16.5) in such a way that the first term on the right-hand side is the result of neglecting the correlations which may develop in time between the moving particle and the scatterers. The second term takes these correlations into account. In order to get an expression for the second term, one has to multiply (16.3) by $w(z|x)$ and average again. This leads to an equation that involves correlations between two scatterers and the moving particle. Thus one has a set of BBGKY hierarchy equations for this and related lattice-gas models.

If we neglect the correlation term on the right-hand side of (16.5) and keep only the first term, we obtain the Boltzmann equation approximation for this model,

$$f(\mathbf{r},\mathbf{c}_i,t+1) = \sum_{\mathbf{r}',\mathbf{c}_j} \delta(\mathbf{r}-\mathbf{c}_i,\mathbf{r}')[\rho W_{ij}+(1-\rho)\delta_{ij}]f(\mathbf{r}',\mathbf{c}_j,t). \tag{16.7}$$

This is a linear equation in f and can be solved for a variety of boundary conditions. Unlike the Boltzmann equation for hard disks that we studied in Chapter 2, this Boltzmann equation leads to a prediction for the diffusion coefficient of the moving particle which is wrong even at low densities, or, at best, only correct for times on the order of a few mean free times! The correlations between the particle and the scatterers are formally of higher order in the density. However, when these correlations are calculated in detail, they actually modify the low-density Boltzmann results.

This modification can be understood in terms of the divergences that appear when one attempts to expand the diffusion coefficient about the Boltzmann equation value in a power series in the density of scatterers. This density expansion must be renormalized by summing the most divergent terms in the expansion. Divergences also occur in the continuous scattering case as well. For the continuous case, the divergences produce modifications to the Boltzmann equation results for appropriate one-dimensional systems, but in two and three dimensions only affect the higher-order density terms, beyond the Boltzmann equation results. For LLG systems, due to the facts that the scatterers are situated on a lattice, and the particle can move only in a few fixed directions, the renormalized diffusion coefficient in this case modifies the Boltzmann equation results in all dimensions. We refer to the papers of Ernst and co-workers, Lebowitz and Percus, and Dorfman and van Beijeren, given below for more details.

16.3 The thermodynamic formalism

Now we turn to the calculation of the chaotic properties of the one-dimensional LLG model. We will show how the thermodynamic formalism defined earlier for deterministic systems can be developed for a stochastic system like this one, as well. This is a rather important observation, for it shows that any stochastic process defined by Chapman–Kolmogorov equations, or by a Markov process, can be classified as a *chaotic system* with most of the properties of a deterministic, chaotic system. †

Our starting point is the dynamical partition function defined on the phase space whose points, $\Omega(t)$, are trajectories of t time steps,

$$Z_L(\beta, t|x_0) = \sum_{\Omega} [\Pi(\Omega, t|x_0)]^{\beta}, \tag{16.8}$$

where $\Pi(\Omega, t|x_0)$ is the probability that the particle follows the trajectory $\Omega(t) = x_1, x_2, \ldots, x_t$ starting at x_0 at $t = 0$, on a lattice of L sites. As usual, β is the inverse-temperature-like parameter,

† It is hard to identify individual Lyapunov exponents for a Markov chain, but a quantity that can be identified with sums of positive Lyapunov exponents can be defined. For more details see Ref. [GD95]

whose variation allows us to scan the structure of the probability distribution Π. In terms of the transition matrix $w(x|y)$, Π is defined, for a fixed configuration of scatterers, by

$$\Pi(\Omega, t|x_0) = \prod_{n=1}^{t} w(x_n|x_{n-1}). \tag{16.9}$$

The partition function requires a calculation of the matrix $w_\beta(x|y)$, which is defined by raising each matrix element of $w(x|y)$ to the power β. It is important to realize that $w_\beta(x|y)$ is a very large matrix of size $2L \times 2L$, which encodes the location of each scatterer on the line, and the transition probabilities for forward and backward scattering at each site. For a fixed configuration of scatterers, the dynamical partition function, $Z_L(\beta, t|x_0)$ can be expressed in terms of the eigenvalues, $\Lambda_{L,i}(\beta)$, and eigenvectors of the matrix $w_\beta(x|y)$. The topological pressure $P(\beta)$, for the fixed configuration, is then equal to the logarithm of the largest eigenvalue of $w_\beta(x|y)$, assuming that such a largest eigenvalue exists and is non-degenerate.†

We are interested in the average topological pressure which is given by

$$\begin{aligned} \psi_L(\beta, \rho) &= \lim_{t\to\infty} \frac{1}{t} \langle \ln Z_L(\beta, t|x_0) \rangle \\ &= \langle \Lambda_{L,\max}(\beta) \rangle , \end{aligned} \tag{16.10}$$

where we average over all configurations of scatterers such that at each site, independently, ρ is the probability that a scatterer will be found there. We also have expressed the average topological pressure in terms of the largest eigenvalue of the matrix w_β.

At this point, a calculation of $\psi_L(\beta, \rho)$ becomes quite complicated. Careful studies have shown that in the limit $L \to \infty$, with fixed density, the topological pressure has two branches, a 'low

† Actually, there is a slight complication here due to the fact that this system consists of two ergodic components. This follows from the observation that w_β is a periodic matrix of period two due to the fact that the particle moves from even numbered sites to even numbered sites, or from odd numbered sites to odd numbered sites, every two steps. Thus the particles on odd sites form one ergodic component, and the particles on even sites form another component. One can restrict oneself to either component by always working with the square of the matrix instead of the matrix itself, and by redefining a new time step to be twice the original time step. We ignore this complication here.

temperature' branch and a 'high temperature' branch with a slope discontinuity at $\beta = 1$. These branches are caused by rare fluctuations in the density of scatterers leading to a localization of the particle in rare, but large empty regions for $\beta > 1$ and in rare, but large filled regions, for $\beta < 1$. The slope discontinuity in $\psi(\beta, \rho)$ is an example of a *dynamical phase transition* in which a singularity appears in the topological pressure or in one of its derivatives.

Here we show how to obtain a mean-field approximation to the topological pressure, for a system with periodic boundary conditions. We do this to illustrate some interesting methods, even though, in view of the above remarks, the mean-field result does not describe the rich behavior of the actual topological pressure.

The mean-field approximation consists of replacing the actual matrices $w_\beta(x_n|x_{n-1})$ below (16.9) by their averages taken over an ensemble of scatterers, as described above. For a system with periodic boundary conditions, this average matrix is organized as follows: the rows and columns are listed in order of lattice site, with two rows or columns for each site corresponding to particles moving to the right ($\mathbf{c} = +1$) and to the left ($\mathbf{c} = -1$), respectively. For instance, a particle arriving at site j with a velocity to the right must have come from site $j-1$ where it either was not scattered at all, scattered in the forward direction, or was reflected by a scatterer at that site. Similarly, a particle arriving at site j moving to the left must have come from site $j+1$. The periodic boundary conditions imply that site L is to the left of site 1. Then the average of w_β, denoted by $\bar{w}_\beta$, is

$$\bar{w}_\beta = \begin{bmatrix} 0 & 0 & 0 & 0 & 0 & 0 & \cdots & a & b \\ 0 & 0 & b & a & 0 & 0 & \cdots & 0 & 0 \\ a & b & 0 & 0 & 0 & 0 & \cdots & 0 & 0 \\ 0 & 0 & 0 & 0 & b & a & \cdots & 0 & 0 \\ \vdots & \vdots & \vdots & \vdots & \vdots & \vdots & \cdots & \vdots & \vdots \\ b & a & 0 & 0 & 0 & 0 & \cdots & 0 & 0 \end{bmatrix}. \tag{16.11}$$

Here $a = 1 + \rho(p^\beta - 1)$ and $b = \rho q^\beta$.

The reader will notice that the matrix on the right-hand side of (16.11) has a rather simple structure. We need to obtain its eigenvalues, the largest one in particular. That is, we want to solve the equation

$$\bar{w}_\beta u = \Lambda u, \tag{16.12}$$

where u is a column vector whose elements are ordered as $u = [u_+(1), u_-(1), u_+(2), u_-(2), \ldots, u_+(L), u_-(L)]^T$, where the superscript T denotes a transpose.

The eigenvalue equations are then

$$\begin{aligned} \Lambda u_+(l) &= a u_+(l-1) + b u_-(l-1) \\ \Lambda u_-(l) &= a u_-(l+1) + b u_+(l+1), \end{aligned} \tag{16.13}$$

for $l = 1, 2, \ldots, L$, and the periodic boundary conditions require that site $L+1 \equiv 1$ and site $0 \equiv L$. We look for solutions of the form $u_\pm(l) = A_\pm \exp(ikl)$ for some wave number k. By setting the resulting determinant in (16.13) equal to zero, we obtain

$$\Lambda(k) = a \cos k + [b^2 - a^2 \sin^2 k]^{1/2}. \tag{16.14}$$

The boundary conditions require that $k = 2n\pi/L$ for $n = 0, 1, \ldots, L-1$. The largest eigenvalue is then $\Lambda_{L,\max}(\beta) = (a + b)$. Thus, in the mean-field approximation one finds that the average topological pressure is

$$\psi(\beta, \rho) = \ln\left[1 + \rho(p^\beta + q^\beta - 1)\right]. \tag{16.15}$$

We leave it to the reader to calculate the Lyapunov exponent, h_{KS}, and h_{top} in the mean-field approximation. We do note that the topological pressure vanishes at $\beta = 1$, as it should.

16.4 Further reading

A good introduction to the statistical mechanics of cellular automata lattice gases is to be found in the article by Ernst [Ern91]. The kinetic theory of the stochastic Lorentz lattice gas is discussed in a series of papers by van Velzen *et al.* [vVE87, vVED88, EvVB89, EvV89b, EvV89a] and in his Ph.D. thesis [vV90]. The application of chaotic dynamics to a study of transport in this model is given by Ernst *et al.* [DEJ95, EDNJ95], and a full discussion of the role of large fluctuations producing dynamical phase transitions is given by Appert *et al.* [AvBED96, AvBED97, AE97]. An extensive discussion of hydrodynamic processes in stochastic lattice gases, with many references, can be found in the book by Spohn [Spo91].

A discussion of the divergences in the density expansions of transport coefficients for hard-disk or hard-sphere systems is given by Dorfman and van Beijeren [DvB77], and for hard-rod systems, by Lebowitz and Percus [LP67, LPS68]. The renormalization of the divergences in the density expansion of transport coefficients results in the theory for the long-time tails of the Green–Kubo time–correlation functions. See [DvB77, PR75, Coh93] for further details.

16.5 Exercises

1. Use the Boltzmann equation for the one-dimensional Lorentz lattice gas, given by (16.7) to compute the diffusion coefficient for this model. Compare your result with the exact value given by $D = p/(2q\rho)$.

2. Determine the chaotic properties of the one-dimensional LLG using the mean-field expression for the topological pressure given by (16.15).

17

Dynamical foundations of the Boltzmann equation

We can now assemble many but, as we shall see, not all, of the pieces we need to construct a consistent picture of the dynamical foundations of the Boltzmann equation and similar stochastic equations used to describe the approach to equilibrium of a fluid or other thermodynamic system. While there are many fundamental points which still are in need of clarification and understanding, our study of hyperbolic systems with few degrees of freedom has pointed us in some interesting directions. In earlier chapters, we saw that the baker's transformation is ergodic and mixing. Moreover, when one defines a distribution function in the unstable direction, one obtains a 'Boltzmann-like' equation with an H-theorem. That is, there exists an entropy function which changes monotonically in time until the distribution function reaches its equilibrium value, provided the initial distribution is sufficiently well behaved, e.g., not concentrated on periodic points of the system. Moreover, the approach to equilibrium takes place on a time-scale which is determined by the positive Lyapunov exponent and is typically shorter than the time needed for the full phase-space distribution function function to be uniformly distributed over the phase-space. Although we can make all of this clear for the baker's transformation it is not so easy to reproduce these arguments in any detail for realistic systems of physical interest. However, we can study more complicated hyperbolic maps to isolate the features we expect to use in a deeper discussion of the Boltzmann equation itself. We will see here that hyperbolic systems in general, even those of few degrees of freedom, exhibit the features that we need to understand the origin of stochastic-like behavior in deterministic mechanical systems, such as a gas or liquid, to which the Boltzmann equation or its higher-density versions are applied. The assumption of hyperbolicity is crucial for the arguments here, although most systems we treat in the laboratory may not actually be hyperbolic.† Nevertheless, we hope that at some time in the future, the main features of this discussion will be shown to apply to systems with some non-hyperbolic properties, such as Lennard–Jones-like particles which are known to have orbiting two-body motions.

17.1 The Arnold cat map

One of the unsatisfactory features of our discussion of the irreversible behavior of the baker's transformation is that we can easily project the full distribution on the unit square onto the unstable direction. One might ask: What would happen if we were to project the full distribution function onto an arbitrary direction? Would one then obtain a Boltzmann-like equation with an H-theorem and an equilibrium distribution? An affirmative answer seems likely, at least for an Anosov system. For if one projects the distribution function onto some arbitrary direction, one will still have a spreading and smoothing of points along some of the unstable directions, and this stretching and smoothing should lead to the usual irreversible behavior, with a projected distribution function approaching its equilibrium value on a time-scale determined by the positive Lyapunov exponents. To see how this might work in general, we return to another class of Anosov, measure-preserving systems of low dimensionality, – the toral automorphisms – and the Arnold cat map in particular.

We recall that the cat map is a hyperbolic toral automorphism defined by the matrix T_A

$$T_A = \begin{bmatrix} 2 & 1 \\ 1 & 1 \end{bmatrix}. \tag{17.1}$$

The cat map has one positive and one negative Lyapunov exponent, and stable and unstable directions which are orthogonal to each other and aligned at a non-zero angle with respect to the Cartesian axes of the coordinate system in which the representation of T_A above is obtained. In Sec. 8.4 of Chapter 8, we have

† They certainly are not classical mechanical systems either. For this reason, among others, a quantum mechanical version of this discussion is highly desirable.

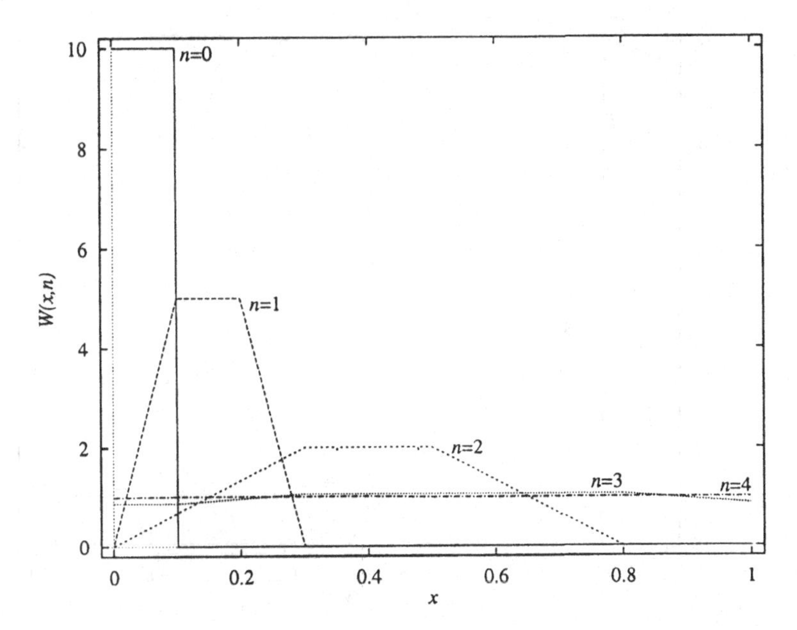

Fig. 17.1 The projection, $W(x,n)$, onto the x-coordinate, of the phase-space distribution function for the Arnold cat map.

illustrated the time evolution of a set of points initially confined to a corner of the unit square. There we saw that after about ten iterations of the map, the points are distributed quite uniformly over the unit square. Although these points must lie on lines oriented along the unstable direction of the map, these lines are so closely and uniformly spaced that the distribution of the points appears to be macroscopically uniform.

If we were to project the phase-space distribution function onto the x- or the y- axes we would be projecting the distribution function onto directions that make an angle with respect to the unstable direction that is neither zero nor $\pi/2$. It would then be interesting to examine the time development of these projected distribution functions. The two-dimensional distributions are illustrated in Figs. 8.3–8.6. In Fig. 17.1 we show the function, $W(x,n)$, obtained by projecting the phase-space distribution function onto the x-axis, at a number of time steps, denoted by n. We see that initially, $W(x,n)$ is non-zero only in a small region around $x = 0$, but as the number of iterations increases, this distribution be-

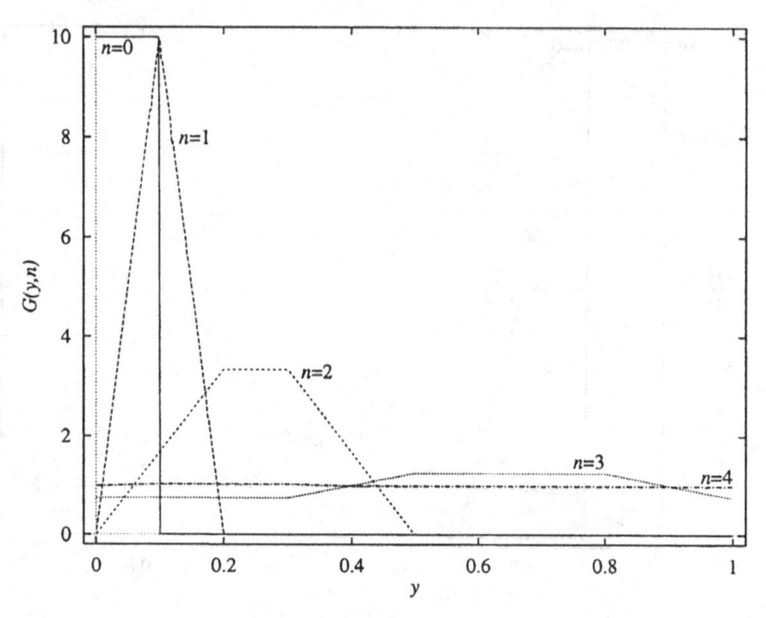

Fig. 17.2 The projection, $G(y, n)$, onto the y-coordinate, of the phase-space distribution function for the Arnold cat map.

comes smoother and smoother, eventually reaching a uniform distribution along the x interval $[0, 1]$. Similarly, we can project the phase-space distribution function along the y-axis. In Fig. 17.2 we show that the projected distribution, $G(y, n)$, starts from a very non-uniform distribution in y, but that after some iterations of the map, it too becomes uniform along the y interval $[0, 1]$. Notice that after three or four time steps, the projected distribution functions have reached near equilibrium values, but the phase-space distribution, as seen in Fig. 8.5, is still far from uniform! This example certainly suggests that the relevant time-scales for the approach to equilibrium of a typical system is determined by the inverses of the positive Lyapunov exponents, which as we see in the next chapter, are usually on the order of a mean free time, with logarithmic adjustments, rather than by the mixing time for the system in the full phase-space. It is important to note that we chose an initial ensemble composed of a very large number of points. The projected distribution function rather quickly approached its equilibrium form. However, if we were to observe only

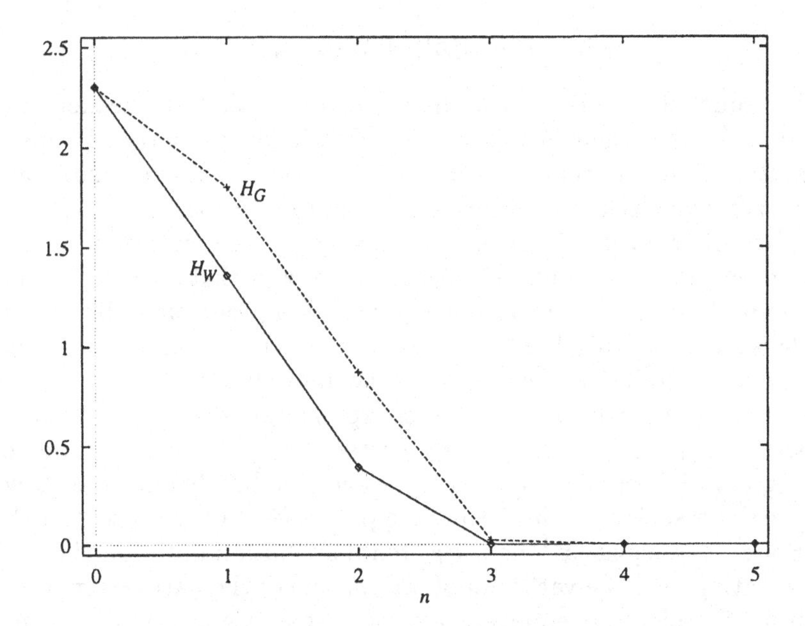

Fig. 17.3 The Boltzmann H-functions $H_W(n)$ and $H_G(n)$, obtained by using $W(x,n)$ and $G(y,n)$ respectively.

a few trajectories, it would take a very much longer time to detect the ergodic behavior of the system, even if we were to project the trajectories on one direction or another.

The computer is a great convenience. We can now ask it to give us the time development of the Boltzmann entropy functions, $H_W(n)$, and $H_G(n)$, given by

$$\begin{aligned} H_W(n) &= \int_0^1 dx W(x,n) \ln[W(x,n)] \\ H_G(n) &= \int_0^1 dy G(y,n) \ln[G(y,n)]. \end{aligned} \tag{17.2}$$

Both of these H-functions decrease steadily with the number of iterations, n, of the cat map, as we would expect and as we observe in Fig. 17.3.

17.2 The Boltzmann equation

We conclude from this example and others that the reader may conclude that there is a general stochastic-like property of Anosov systems, namely that projected distributions will irreversibly approach equilibrium distributions provided that the initial phase-space distribution function is sufficiently well-behaved, i. e., not concentrated on points of measure zero, and that the projections are in directions that are not entirely along the stable directions. This stochastic-like behavior has its origins in the fundamental dynamical instability of hyperbolic systems, whereby a small change in an initial condition leads to an exponential separation of phase-space trajectories. Thus, even systems with a small number of degrees of freedom may display stochastic-like behavior that we usually associate with irreversible processes. In the context of the Boltzmann equation, the projected distribution is obtained by integrating over the variables of all but one of the particles in going from the Liouville equation for the N-particle distribution function to the BBGKY hierarchy equations for the reduced distribution functions.

While our arguments based upon the properties of hyperbolic systems provide us some comfort in understanding how stochastic equations can result from projecting a reversible equation onto an unstable direction, or directions, we still have not clearly understood the dynamical origins of the *Stosszahlansatz*. Put precisely, the question is the following: Does the stochastic-like behavior of a system of a few degrees of freedom have anything at all to do with the stochastic-like behavior of a system composed of a large number of particles, as described by the Boltzmann equation, for example?

It seems clear that the answer to this question is 'yes', and that two fundamental issues are involved in providing this answer. First, we must assume, or prove somehow, that our system of many particles can be thought of as a hyperbolic dynamical system, in so far as its transport properties are concerned. Then we can safely argue that projected phase-space distributions, at least, should exhibit the same kind of irreversible behavior that we see in small hyperbolic systems. However, and this is the second point, the details of irreversible behavior of the system will indeed depend upon its size in a crucial way, as we have argued previously in Chapter 1, and describe in more detail in Section 17.5.

17.3 Stochastic equations

At this point, it is reasonable to ask why one should be so concerned with the microscopic foundations of the theory of irreversible processes. After all, one might say stochastic equations like the Langevin equation, and even the Boltzmann equation itself, are extremely successful in predicting the behavior of a variety of non-equilibrium systems. Of course, it is deeply unsatisfactory to ignore the fact that fluids are composed of particles that obey the laws of mechanics, and to ignore any effects produced by the microscopic and mechanical nature of fluid systems. Such effects are indeed important, but very often even they can be produced by some stochastic arguments.† However, one might also think that something serious will eventually go wrong if one ignores the underlying microscopic structure of a fluid. One such effect was pointed out by Kolmogorov and elaborated upon by Gaspard and Wang. These authors analyzed the rate of entropy production for a number of different types of systems, from purely deterministic to purely stochastic. They generalized the notion of KS entropy to a notion of an (ϵ, τ)-entropy per unit time, where ϵ describes the spatial resolution of some measuring process and τ describes the time-resolution of the same measuring process. If one supposes that stochastic processes remain stochastic on arbitrarily fine spatial and temporal scales, then the rate of entropy production can be argued to be *infinite* in such a process, as $\epsilon \to 0$. On the other hand, the KS entropy of a deterministic, mechanical system remains finite as one reduces the scale of observation to arbitrarily fine levels. However, in view of the very fine spatial scale needed to see the difference between the rate of entropy production in deterministic and stochastic processes, it seems unlikely that one could tell from experimental measurements alone that a system is

† A good example of this is the theory for the long-time tails in the time-correlation functions appearing in the Green–Kubo formulae. These tails can be derived both from a microscopic kinetic theory as well as from arguments based upon adding a Gaussian-white noise term to the Navier–Stokes equations.

a deterministic dynamical system and not a stochastic one. Thus, as one might expect, on fine scales there is a very clear difference between stochastic and deterministic systems. Treating fluid systems as chaotic dynamical systems allows us then to assume a degree of disorder, as measured by the Kolmogorov–Sinai entropy, which is consistent with their microscopic dynamics, and avoids the unsatisfactory consequences of a purely stochastic picture of fluid systems. With this in mind, we turn to a further elaboration of this theme provided by the chaotic hypothesis of Cohen and Gallavotti.

17.4 The chaotic hypothesis

Cohen and Gallavotti have proposed that as a working hypothesis, we treat fluid systems as if they were Anosov hyperbolic systems and explore the consequences of that assumption, before wrestling with the harder problem of systems that have some kind of non-hyperbolic properties. This *chaotic hypothesis* is for non-equilibrium systems, the analog of the ergodic hypothesis, which can be used to justify the calculations of equilibrium statistical mechanics, although we have never proved that a laboratory system is indeed ergodic.

17.5 The thermodynamic limit

Supposing that we agree that a system of 10^{23} particles is an Anosov dynamical system and that projected distribution functions satisfy a stochastic equation, how can we prove that this equation reduces to the Boltzmann transport equation in the dilute gas limit? To construct an answer, we think of an ensemble distribution function that is smooth in some sense in the phase-space. This requires that we consider a very large number of members of an ensemble, and that we construct a normalized probability distribution function in phase-space. Then the hyperbolic nature of the dynamics can take us very far in understanding the apparently irreversible behavior of the reduced distribution functions. However, the details of this irreversible behavior certainly depend on the details of the system, and we can expect that a fundamental proof of the validity of the Boltzmann equation must take into account not only the hyperbolic, or hyperbolic-like, dynamics but also the fact that a gas is composed of 10^{23} or so particles. The use of the laws of large numbers may make issues such as the validity of the *Stosszahlansatz* more transparent in the future.

To understand why the thermodynamic limit plays such a crucial role in our understanding of the foundations of the Boltzmann equation, we consider a model commented upon by a number of authors (see *Further reading*). Imagine an isolated box with a small number, say five, of hard spheres within it, and suppose further that the size of the box is sufficiently small that collisions among the hard spheres are frequent enough so that we can follow the spheres over the course of many collisions. Although we may safely assume that the dynamics is ergodic and mixing, provided the boundary conditions are innocuous, it would still be hard to follow the motion of the system, starting from some given initial state, over a long period of time, and then conclude that the motion is irreversible in any sense. We could, though, imagine a situation where all the five spheres were given an initial kinetic energy, and placed in some *randomly chosen* configuration in one of the corners of the box. Then we can be sure, because of the mixing property of the system, that the spheres would leave the corner and make their way into the rest of the box, and that the spatial distribution of the spheres would eventually fluctuate about a uniform average value. This uniform average value can, of course, be identified with the average behavior of an ensemble of identically prepared systems with slightly different initial states. However, for an individual system of five spheres the fluctuations can be quite large, and the Poincaré recurrence time can be quite short. We would therefore not expect that the laws of irreversible thermodynamics would apply to this one system. We have seen, however, for the baker's map and the cat map, that the laws of irreversible thermodynamics do apply to the average behavior of an ensemble of such systems.†

† An example of this phenomenon can be found in the computer simulation world. There one considers a periodic continuation of a small system. For example, simulations are routinely carried out on systems of two or three particles placed in a small volume and then this system is periodically repeated as a tiling of space. Values for viscosities obtained in such

Returning to our one system with five spheres, we note that due to the large fluctuations we can imagine that the initial state of the spheres, with all of them in one corner, might not be the result of an artificially prepared initial configuration, but rather due to a naturally occurring fluctuation. In this sense, we could not distinguish irreversible behavior from naturally occurring fluctuations on the scale of a laboratory experiment. The use of a randomly chosen initial state is also crucial, since we might imagine initial states where the particles either never get out of the corner, or go into peculiar, periodic states if they do get out. In any case, the role of an *ensemble of initial states* is to provide a generic description of our system, to remove the effects of strange initial states that are not likely to occur in a real laboratory situation, and to enable us to express the average behavior of a large number of systems in simple laws. In this last respect, we have seen that the average behavior of suitably defined properties of a hyperbolic system of even a few degrees of freedom can obey rather simple laws of irreversibility.

However, if the system consisted of 10^{23} hard spheres, the conclusions that we would draw would be very different, if we started from a configuration where most of the spheres were placed in one corner. We wouldn't expect to see all of these spheres, or even a significant fraction of them, reappear in one corner of a (very big) box if the box were left untouched for a long time, nor would we explain such an occurrence as a fluctuation. Instead, we would look around to find the culprit who artificially put the spheres in the corner. Moreover, we would expect that the flow of spheres out of the corner and into the rest of the box would be described by irreversible thermodynamics. We would continue to construct ensemble averages to capture the generic nature of the initial configuration. We would then expect that the thermodynamic limit, from this point of view, will increase the Poincaré recurrence time and suppress large fluctuations. Thus, as the number of particles

simulations are not substantially different from those of systems with large numbers of particles [Hoo91, BS96]. Of course, all computer simulations are subject to random perturbations associated with the round-off error in the computer arithmetic. This leads to interesting questions about the role of random perturbations in the time evolution of dynamical systems. See the discussion in Chapter 19 on this question.

becomes large, we expect the distribution function for the macroscopic variables to become very sharp. This would guarantee that essentially all of the members of the ensemble exhibit the same macroscopic behavior. We might also suppose that it may not even be necessary to show that our system is hyperbolic. Rather, we might only need to prove that the system is hyperbolic in a subspace spanned by the macroscopic variables.

Another natural role of the large-system limit, and one that is no less important for our understanding of the Boltzmann equation, is to provide one of the mechanisms whereby the dynamical correlations between particles can be made very small. That is, at low densities, at least, particles will typically collide with particles with which they have no previous dynamical correlations. Of course, at high densities dynamical correlations can be very important indeed, and are responsible for algebraic decays of time-correlation functions, and other related phenomena.

In summary, we have seen that irreversible behavior is characteristic of all hyperbolic systems, whether of few or many degrees of freedom. We don't expect one *small* system in the laboratory to exhibit apparently irreversible behavior, although it would show irreversible behavior if we were to compute the appropriate time averages by following the motion of the particles. A large system would easily show an irreversible behavior to equilibrium because the fluctuations of an individual system about an average behavior are likely to be small. It appears that both features – the hyperbolicity of the dynamics, and the large-system limit – are necessary for an understanding of the origins of the Boltzmann equation. The hyperbolicity guarantees that an irreversible equation will describe the behavior of the reduced distribution functions, and the large-system limit will provide the mechanism for the correctness of the *Stosszahlansatz* for a dilute gas, and for corrections to it for a dense one.

For systems of hard spheres, Lanford, and later Sinai and coworkers have made progress in providing a firm basis for the Boltzmann equation. For example, Sinai, Boldrighini, and Bunimovich have provided a rigorous derivation of the Boltzmann equation for the Lorentz gas where one particle moves in an array of fixed scatterers, in the appropriate dilute scatterer limit, called the *Grad limit*. Here, the thermodynamic limit and the suppression of the effects of fluctuations are due to the large number of scatterers, and the construction of an average over a large number of initial states of the *one* moving particle. However, the problem posed at the beginning of this book, i.e., how to derive the Boltzmann equation from first principles using the chaotic nature of the dynamics and the large-system limit, is still to be fully resolved.

We conclude this chapter by pointing out the fact that the Boltzmann equation provides a key toward the understanding of the transition between *microscopic* chaos and *macroscopic* chaos. That is, to understand the dynamical foundations of the Boltzmann equation, as we have argued here, one needs to understand the chaotic properties of the interactions between the particles in the fluid. However, granting for the moment the validity of the Boltzmann equation, one can use it to obtain the non-linear Navier–Stokes equations of fluid dynamics. It is generally agreed that these equations have solutions, if one can call them that, which exhibit turbulent behavior, characteristic of some kinds of macroscopic chaos. Thus, the behavior of turbulent fluids is coded in some way in the elementary but chaotic dynamics of the particles that compose the fluid.

17.6 Further reading

Many of the issues discussed here are also treated in the papers by Dorfman and van Beijeren [DvB97, Dor98]. A treatment of the rate of entropy production, the (ϵ, τ)-entropy per unit time, for various kinds of processes, from deterministic to stochastic, together with references to the papers of Kolmogorov, can be found in Gaspard's book [Gas98], as well as in the paper of Gaspard and Wang [GW93]. The chaotic hypothesis is discussed in papers by Gallavotti and Cohen [GC95a, GC95b]. The work on hard-disk and hard-sphere systems is described in the collections of papers edited by Sinai [Sin89, Sin91], and the derivation of the Boltzmann equation for the Lorentz gas is given in Boldrighini *et al.* [BBS83]. The reader is also advised to read the book by Krylov, edited by Sinai, and to read Sinai's beautiful appendix in that volume [Kry79]. Lanford's derivation of the Boltzmann equation is found in [Lan76, Lan81].

The discussion of the role of chaos compared to the role of the thermodynamic limit is taken up in papers by Lebowitz [Leb93], Gallavotti [Gal95], by Bricmont [Bri95], and by Cohen and Rondoni in the special issue of *Chaos* on *Chaos and Irreversibility* [TGN98]. The paper of Bricmont discusses additional topics, and the debate between Prigogine and Bricmont in the same issue of *Physicalia* is instructive. The reader will also be amused by the recent computer study by Bowles and Speedy which followed five disks in a box over several hundred thousand collisions, and demonstrated that systems with as few as five particles show many of the features of much larger systems, such as fluid, crystalline, and glass phases [BS99].

There is an interesting literature on decays of time–correlation functions in dynamical systems. Some good references in the mathematical literature are the papers by Liverani [Liv95] and by Young [You99]. The decay of the time–correlation functions appearing in the Green–Kubo formulae for fluids are described in the articles by Dorfman and van Beijeren [DvB77], by Pomeau and Resibois [PR75], and by Dorfman, Kirkpatrick, and Sengers [DKS94], among many others.

18
The Boltzmann equation returns

We began this excursion into the dynamical systems approach to nonequilibrium statistical mechanics with a discussion of the Boltzmann transport equation. We end this excursion with the Boltzmann equation, but now we are going to use it to compute some Lyapunov exponents. The fact that the Boltzmann equation begins and ends this book may serve to illustrate both the power and the beauty of this equation, sitting at the heart of our understanding of irreversible phenomena.

18.1 The Lorentz gas as a billiard system

We are going to calculate the positive Lyapunov exponent for a two-dimensional hard-disk Lorentz gas. To do so, we will combine ideas of Boltzmann with those of Sinai, thus completing, in some sense, the transition from molecular chaos to dynamical chaos, and showing the deep connection between them.

Imagine then a collection of hard disks of radius a placed at random in the plane at low density, i.e., $na^2 \ll 1$, where n is the number density of the disks (see Fig. 18.1). Next, imagine a point particle moving with speed v in this array. The particle moves freely between collisions with the disks and makes specular collisions with the disks from time to time, preserving its speed and energy, but not its momentum upon collision. Sinai has considered some of the mathematical properties of this system, and has proved that it is mixing and ergodic. The moving particle has four degrees of freedom – two coordinates and two momenta – but the energy is conserved. Therefore, the phase-space of the moving particle is three-dimensional. There is a possible total of three Lyapunov exponents for the phase-space trajectories in this three-dimensional phase-space. One of these Lyapunov exponents

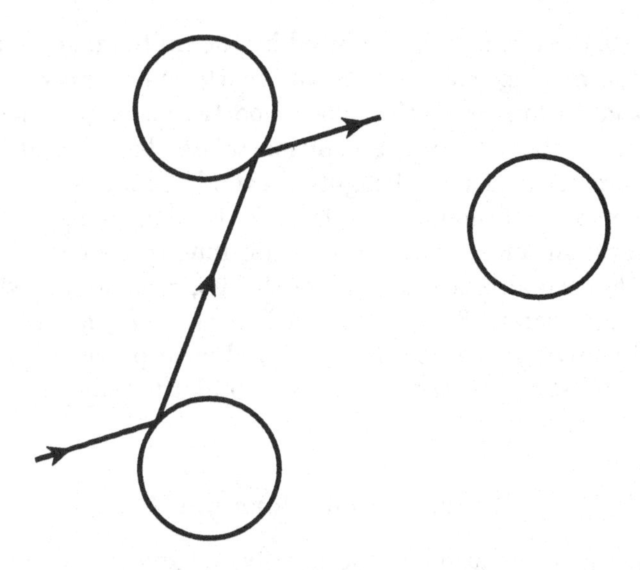

Fig. 18.1. The two-dimensional Lorentz gas.

is zero and corresponds to the direction which is tangent to the phase-space trajectory. That is, two phase points that lie close to each other on the same trajectory do not separate exponentially. For the Lorentz gas this is quite obvious, because the two trajectories maintain their initial separation for all time since the speed of the particle stays constant. In more general cases, the non-exponential separation of two phase points on the same trajectory is a bit harder to prove.†

This leaves two directions, perpendicular to the trajectory, with non-zero Lyapunov exponents. Since this is a Hamiltonian, symplectic system, the non-zero Lyapunov exponents come in $\pm$ pairs, as in the baker's transformation. Here we calculate the positive Lyapunov exponent; the negative one follows from this pairing rule.

It is important to note that this system is not one of the usual hyperbolic systems that we have been studying. Instead of having nice continuous stable and unstable manifolds, this billiard

† One method uses the Poincaré recurrence theorem, and the reader is referred to the *Reviews of Modern Physics* article by Ruelle and Eckmann for a detailed explanation.

system has discontinuities produced by the finite range of interaction of the moving particle with the scatterers. Simply stated, if the impact parameter of the moving particle is slightly less than the radius of the scatterer, a collision takes place, but if the impact parameter is increased slightly beyond the radius of the scatterer, then no collision takes place. This phenomenon makes a billiard system an example of a discontinuous system with non-uniform hyperbolic behavior. Nevertheless, Sinai and co-workers, Gallavotti, Ornstein, Katok, Stelcyn, Liverani, Wojtkowski, Krámli, Szász, Simányi, and others have been able to prove a number of ergodic, and stronger, theorems about billiard systems.

18.2 Sinai's formula

To begin the calculation of the positive Lyapunov exponent, we consider a typical point on the trajectory of the particle when it is in between two collisions with the scatterers. We want to compute the rate of separation of this trajectory from one close to it, in phase-space. Let's pick the nearby trajectory to be one which is at the same spatial point, but with a velocity direction which differs very slightly from that of our reference trajectory. We denote the reference trajectory by $\Gamma(t) = (\mathbf{r}(t), \mathbf{v}(t))$ and the nearby trajectory by $\Gamma(t)+\delta\Gamma(t) = (\mathbf{r}(t)+\delta\mathbf{r}(t), \mathbf{v}(t)+\delta\mathbf{v}(t))$. The fact that the energy is the same for the moving particle on both trajectories means that $\mathbf{v}(t)\cdot\delta\mathbf{v}(t) = 0$. Now let's focus on the geometry of these two trajectories from the initial instant where they begin to separate. Think of the two trajectories as being so close together that the two trajectories involve collisions with the same scatterers for an arbitrarily large number of collisions. This corresponds, of course, to finding the Lyapunov exponents for a map by picking two points an infinitesimal distance apart. If we follow the two trajectories for a time t_0 before the first collision, we see that they form a small triangle, or equivalently, a wedge of a circle with arc length $\delta S(t_0) = \rho(t_0)\delta\theta$, and radius $\rho(t_0) = vt_0$, where $\delta\theta$ is the angle between $\mathbf{v}$ and $\mathbf{v}+\delta\mathbf{v}$ (see Fig. 18.2). We take t_0 to be some arbitrary time before the first collision with a scatterer, and we take the first such collision to occur at time $t_0+\tau_1$. The arc length of the circular wedge then becomes

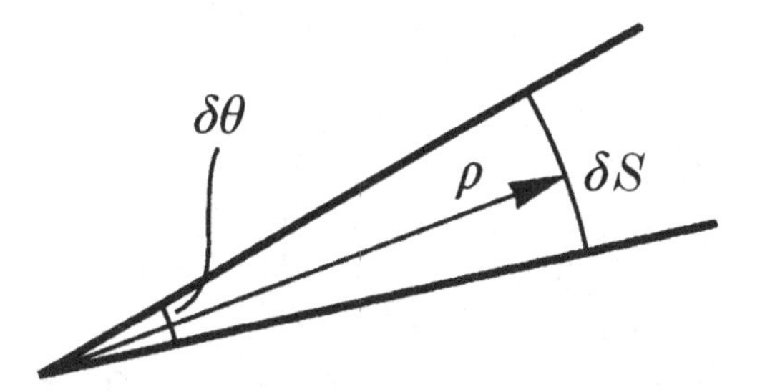

Fig. 18.2 The growth of the radius of curvature between collisions.

$$\delta S(t_0 + \tau_1) = v(t_0 + \tau_1)\delta\theta = \delta S(t_0)\left(1 + \frac{v\tau_1}{\rho(t_0)}\right). \tag{18.1}$$

Here we choose a somewhat awkward notation for later convenience. The collisions of the particle with the scatterers are instantaneous, and the scatterers are circles of radius a. If we think of the little wedge as a pencil of light, then the collision of the two nearby trajectories with the scatterer is identical to the reflection of the pencil of light by a circular mirror! That is, we are dealing with an optical problem. We have a pencil of light with radius of curvature $\rho(t_0 + \tau_1)$ and small angular width impinging on a mirror with radius a. The formulae of classical optics tell us that the radius of curvature after collision, denoted as ρ_+, is related to that before collision, denoted as ρ_-, by

$$\frac{1}{\rho_+} = \frac{1}{\rho_-} + \frac{2}{a\cos\phi}, \tag{18.2}$$

where ϕ is the angle of incidence of the pencil on the disk (see Fig. 18.3). This is a standard formula of elementary optics (usually for $\phi = 0$, for the locations of objects, ρ_-, and images, ρ_+, in front of spherical mirrors), and the proof of the general result is left as an exercise for the reader (see Exercise 18.1 and the references given below).

Let $\rho_{+,1}$ be the radius of curvature after the first collision of the moving particle with a scatterer. It then follows from the above argument that if τ_2 is the time between the first and second collisions, then the separation of the two trajectories just before the

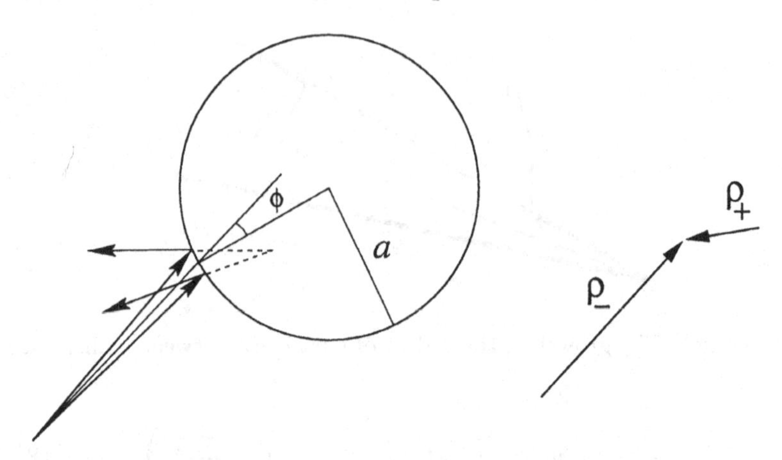

Fig. 18.3. The change in the radius of curvature due to collision.

second collision will be

$$\delta S(t_0 + \tau_1 + \tau_2) = \delta S(t_0)\left(1 + \frac{v\tau_1}{\rho(t_0)}\right)\left(1 + \frac{v\tau_2}{\rho_{+,1}}\right). \tag{18.3}$$

We have used the fact that, although the radius of curvature changes discontinuously at a collision, the separation of the two trajectories is the same just after a collision as just before it, i. e., the separation of trajectories is a continuous function of time. Clearly, if we have a sequence of n collisions, taking place with time intervals $\tau_1, \tau_2, \tau_3, \ldots, \tau_n$ after the initial time t_0, then the separation of the trajectories at some time τ after the nth collision but before the $(n+1)$th collision will be

$$\begin{aligned}&\delta S(t_0 + \tau_1 + \tau_2 + \cdots + \tau_n + \tau) \\ &= \delta S(t_0)\left\{\prod_{i=1}^{n}\left(1 + \frac{v\tau_i}{\rho_{+,i-1}}\right)\right\}\left(1 + \frac{v\tau}{\rho_{+,n}}\right). \end{aligned}\tag{18.4}$$

We can now express this separation of trajectories in exponential form, preparatory to getting a final expression for the positive Lyapunov exponent, as

$$\frac{\delta S(t_0 + T)}{\delta S(t_0)} = \exp T\left[\frac{1}{T}\left\{\sum_{i=1}^{n}\ln\left(1 + \frac{v\tau_i}{\rho_{+,i-1}}\right) + \ln\left(1 + \frac{v\tau}{\rho_{+,n}}\right)\right\}\right] \tag{18.5}$$

where $T + t_0 = t_0 + \tau_1 + \cdots + \tau_n + \tau$. We are now going to argue that the quantity in the square brackets in the exponential expression in (18.5) is well-behaved as $T \to \infty$, so that for large T the right-hand side of (18.5) has the form $\exp(\lambda T)$. We then will identify λ as the positive Lyapunov exponent for this system. To see this we use a simple identity:

$$\ln\left(1 + \frac{vt}{\rho}\right) = v \int_0^t \frac{1}{\rho + v\tau} d\tau.$$

This identity allows us to write

$$\sum_{i=1}^{n} \ln\left(1 + \frac{v\tau_i}{\rho_{+,i-1}}\right) + \ln\left(1 + \frac{v\tau}{\rho_{+,n}}\right)$$
$$= \sum_{i=1}^{n} v \int_0^{\tau_i} \frac{1}{\rho_{+,i-1} + vt} dt + v \int_0^{\tau} \frac{1}{\rho_{+,n} + vt} dt. \quad (18.6)$$

Notice that the denominator in each of the integrals appearing above is simply the radius of curvature for all times in the interval between the $(i-1)$th collision and the ith collision, except for the last integral where it is simply the radius of curvature in the interval after the last collision in the sequence and some next collision, which occurs after time T. Consequently, the right-hand side of (18.6) can be written as one integral

$$v \int_{t_0}^{t_0+T} dt \frac{1}{\rho(t)},$$

and the expression for the separation of trajectories, (18.5), becomes

$$\frac{\delta S(T)}{\delta S(t_0)} = \exp T \left[\frac{v}{T} \int_{t_0}^{t_0+T} \frac{1}{\rho(t)} dt\right]. \quad (18.7)$$

This result allows us to obtain the Lyapunov exponent as the *time average* of the inverse of the radius of curvature,

$$\lambda = \lim_{T \to \infty} \frac{v}{T} \int_{t_0}^{t_0+T} \frac{1}{\rho(t)} dt. \quad (18.8)$$

Here we have taken the limit – suggested by our earlier study of maps – that the two nearby trajectories start infinitesimally close to each other and we let the time interval grow large in order to extract a quantity representing the exponential rate of separation of very nearby trajectories. One might question our identification of this expression with the Lyapunov exponent since we have just examined the *spatial* separation of two trajectories, while the Lyapunov exponent characterizes the total separation of trajectories in *phase-space.* A simple argument shows that these two quantities are the same. To see this, let's look at the separation of trajectories in phase-space

$$\delta\Gamma(t) = \left[\left(\frac{\delta\mathbf{r}(t)}{r_0}\right)^2 + \left(\frac{\delta\mathbf{v}(t)}{v_0}\right)^2\right]^{1/2}. \tag{18.9}$$

Here, r_0 and v_0 are scaling constants chosen so that the dimensions of the two terms in the above expression will be the same, e.g., dimensionless. Now the quantity $|\delta\mathbf{r}(t)|$ is just $\delta S(t)$ computed above. The quantity $\delta\mathbf{v}(t)$ is simply the time derivative of $\delta\mathbf{r}(t)$, so the time derivative of $\delta\mathbf{r}(t)$ will also be an exponential function of time, with the same exponential behavior. Thus both terms in the square root on the right-hand side of (18.9) will grow with the same exponent, λ.

Equation (18.8) is the beautiful formula of Sinai. In the next section, we will see how it can be evaluated for a dilute, random Lorentz gas, using the Boltzmann equation. Before doing so, we recall the individual ergodic theorem of Birkhoff, which asserts that for ergodic systems one may replace time averages by ensemble averages. If one can prove that the random Lorentz gas is ergodic then an expression for λ is given by

$$\lambda = v\left\langle\frac{1}{\rho}\right\rangle_{\text{eq}}, \tag{18.10}$$

where the angular brackets denote an average over an equilibrium ensemble. The proof that the random Lorentz gas is ergodic was given by Sinai – it is quite difficult – but we may conclude from it that (18.10) may be used to compute the Lyapunov exponent.

18.3 The extended Lorentz–Boltzmann equation

Now we are in a position to combine almost everything we have discussed in this book in order to compute the Lyapunov exponent for a dilute random Lorentz gas. We know from general procedures in statistical mechanics that the evaluation of an ensemble average, such as that in (18.10), requires the construction of a distribution function for the ensemble. Further, for a system of one particle moving in an array of scatterers, we need, in general, the combined distribution function for the moving particle and the scatterers in the combined phase-space, consisting of the product of the position and momentum space for the particle, and the configuration space for the fixed scatterers.

Our approximation here will be appropriate for a dilute collection of scatterers. That is, we assume that the moving particle is not correlated with any of the scatterers *before* it collides with them. That is, the distribution for the moving particle is independent of the distribution of the scatterers. Examples of similar systems that we have encountered so far include the Boltzmann equation treatment of the Kac ring model (Sec. 2.3), the Ehrenfest wind-tree model (Exercise 2.1), and the Lorentz gas (Exercise 2.2). We will suppose then that the expression for the Lyapunov exponent can be evaluated by finding an appropriate single-particle distribution function for the moving particle. It is worth pointing out that to go to higher densities, we would need a combined distribution function for the particle and the scatterers.

We will not be able to use the simple distribution functions discussed in Chapter 2 since we have another variable to worry about, namely the radius of curvature ρ. This variable can be incorporated into the equation by making the following construction: The moving particle is described by its position $\mathbf{r}$, its velocity $\mathbf{v}$ and by a radius of curvature ρ. This variable is added by supposing that to each trajectory of the moving particle there is an associated neighboring trajectory which initially is infinitesimally close to the one we are considering. Then the quantity ρ characterizes the separation of the two trajectories by the radius of curvature for the two trajectories - the *reference* trajectory of the moving particle and its fictitious *neighboring* trajectory. We know that the radius of curvature grows linearly in time between collisions,

$$\rho(t) = \rho(t_0) + v(t - t_0) \quad t \geq t_0, \tag{18.11}$$

and changes discontinuously at a collision according to (18.2). This is just what we need for a Boltzmann equation. We consider a distribution function $f(\mathbf{r}, \mathbf{v}, \rho, t)$ and apply the heuristic method of Boltzmann to derive an equation for it.

To proceed, we let $f(\mathbf{r}, \mathbf{v}, \rho, t)d\mathbf{r}d\mathbf{v}d\rho$ be the probability of finding the moving particle with position in the region $d\mathbf{r}$ about $\mathbf{r}$, with velocity in the region $d\mathbf{v}$ about $\mathbf{v}$, and with radius of curvature (with a neighboring trajectory as described above) in the region $d\rho$ about ρ, all at time t. Then, following the discussion in Chapter 2 and in Exercise 2.2, we consider that f changes with time due to the free motion of particles between collisions and due to collisions. Then we write an equation for f as

$$[f(\mathbf{r}+\mathbf{v}dt, \mathbf{v}, \rho + vdt, t+dt) - f(\mathbf{r}, \mathbf{v}, \rho, t)]d\mathbf{r}d\mathbf{v}d\rho = \Gamma_+ - \Gamma_-. \tag{18.12}$$

The Taylor expansion of the left-hand side of (18.12) is immediate, of course, and as in Chapter 2, Γ_- can easily be computed just by considering the collision cylinders in which particles with $\mathbf{v}$ and ρ are lost. That is,

$$\begin{aligned}\Gamma_- &= navf(\mathbf{r}, \mathbf{v}, \rho, t)\int_{-\pi/2}^{\pi/2} d\phi|\cos\phi|d\mathbf{r}d\mathbf{v}d\rho dt \\ &= 2navf(\mathbf{r}, \mathbf{v}, \rho, t)d\mathbf{r}d\mathbf{v}d\rho dt, \end{aligned} \tag{18.13}$$

where ϕ is the angle of incidence in the loss collisions. Remember that the collisions under discussion here are those of the moving particle with fixed scatterers.

We can compute Γ_+ in a very similar way. Here we need to consider restituting collisions that are so constructed that a particle with velocity $\mathbf{v}'$ and radius of curvature ρ' *before collision with a scatterer* produces a particle with velocity $\mathbf{v}$ and radius of curvature ρ *after collision.* The probability per unit volume of finding a particle with velocity $\mathbf{v}'$ and radius of curvature ρ' in small regions $d\mathbf{v}'$ and $d\rho'$ is given by $f(\mathbf{r}, \mathbf{v}', \rho', t)d\mathbf{v}'d\rho'$, and the probability of finding such a particle in a collision cylinder about a scatterer, characterized by angle of incidence in the range ϕ to $\phi + d\phi$ is, as is discussed in Chapter 2,

$$f(\mathbf{r}, \mathbf{v}', \rho', t)d\mathbf{v}'d\rho' nav' \cos\phi\, d\phi dt.$$

Here, the restituting velocity is given by the usual expression $\mathbf{v}' = \mathbf{v} - 2\left(\mathbf{v}\cdot\hat{\mathbf{k}}\right)\hat{\mathbf{k}}$, and $\hat{\mathbf{k}}$ is the unit vector in the direction from the center of the scatterer to the point of incidence of the moving particle. We note that $\mathbf{v}'$ has the same magnitude as $\mathbf{v}$, so that the transformation from $\mathbf{v}'$ to $\mathbf{v}$ is just an orthogonal rotation, and $d\mathbf{v}' = d\mathbf{v}$. Also, ρ' is given by

$$\frac{1}{\rho'} = \frac{1}{\rho} - \frac{2}{a\cos\phi}$$

so that

$$d\rho' = \left(\frac{a\cos\phi}{a\cos\phi - 2\rho}\right)^2 d\rho. \tag{18.14}$$

It is important to notice that the radius of curvature after collision must satisfy the inequality

$$\rho \leq \frac{a\cos\phi}{2}. \tag{18.15}$$

We can now find an expression for Γ_+ as

$$\begin{aligned}\Gamma_+ &= nav\int_{-\pi/2}^{\pi/2} d\phi\cos\phi\left(\frac{a\cos\phi}{a\cos\phi - 2\rho}\right)^2 \Theta(a\cos\phi - 2\rho)\times \\ &\qquad f\left(\mathbf{r},\mathbf{v}',\frac{\rho}{1-(2\rho/a\cos\phi)},t\right) d\mathbf{r}d\mathbf{v}d\rho dt.\end{aligned} \tag{18.16}$$

This expression for Γ_+ can be transformed to a slightly simpler looking expression

$$\begin{aligned}\Gamma_+ &= nav\int_{-\pi/2}^{\pi/2} d\phi\int_0^{\infty} d\rho'\delta\left(\rho - \frac{a\cos\phi}{2[1+(a\cos\phi/2\rho')]}\right)\times \\ &\qquad f(\mathbf{r},\mathbf{v}',\rho',t)\, d\mathbf{r}d\mathbf{v}d\rho dt.\end{aligned} \tag{18.17}$$

Now (18.12), (18.13), and (18.17) can be combined to yield an extended Lorentz–Boltzmann equation for the probability distribution $f(r,\mathbf{v},\rho,t)$

$$\begin{aligned}&\frac{\partial f}{\partial t} + \mathbf{v}\cdot\nabla f + v\frac{\partial f}{\partial \rho} \\ &\quad = nav\int_{-\pi/2}^{\pi/2} d\phi\int_0^{\infty} d\rho'\cos\phi\,\delta\left(\rho - \frac{a\cos\phi}{2[1+(a\cos\phi/2\rho')]}\right)\times \\ &\qquad f(\mathbf{r},\mathbf{v}',\rho',t) - 2navf(\mathbf{r},\mathbf{v},\rho,t).\end{aligned} \tag{18.18}$$

This is the equation that can now be used to compute the positive Lyapunov exponent for the dilute Lorentz gas. It is very important to note that if this equation is integrated over all values of the radius of curvature ρ, one then obtains the usual Lorentz–Boltzmann equation

$$\begin{aligned}&\frac{\partial F(\mathbf{r},\mathbf{v},t)}{\partial t} + \mathbf{v}\cdot\nabla F(\mathbf{r},\mathbf{v},t) \\ &\quad = nav\int_{-\pi/2}^{\pi/2} d\phi\cos\phi[F(\mathbf{r},\mathbf{v}',t) - F(\mathbf{r},\mathbf{v},t)],\end{aligned} \tag{18.19}$$

where $F(\mathbf{r},\mathbf{v},t) = \int d\rho f(\mathbf{r},\mathbf{v},\rho,t)$ is the ordinary probability distribution function for a moving particle in the Lorentz gas.

In order to compute λ_+, we will look for equilibrium solutions of (18.18) where f does not depend on position $\mathbf{r}$ or time t. This simplifies the solution a bit, but we can make an even more important simplification by realizing that the typical radius of curvature of the trajectories just before a collision with a scatterer will be on the order of the mean free path between collisions, which is of order $(na)^{-1}$. Then we can neglect the ρ' term in the argument of the Dirac delta function in (18.18), obtaining

$$v\frac{\partial f(\mathbf{v},\rho)}{\partial \rho} = nav\int_{-\pi/2}^{\pi/2} d\phi \cos\phi\, \delta\left(\rho - \frac{a\cos\phi}{2}\right) F(\mathbf{v}') - 2navf(\mathbf{v},\rho), \tag{18.20}$$

where

$$F(\mathbf{v}) = \int_0^\infty d\rho f(\mathbf{v},\rho).$$

Because the gas is in equilibrium, the velocity distribution function $F(\mathbf{v})$ is just a constant, since the speed of the particle is constant, and the possible directions, θ, of motion, are uniformly distributed over the interval $0 \le \theta \le 2\pi$. We may take $F(\mathbf{v}) = (2\pi)^{-1}$ and also notice that f need not be a function of the velocity vector $\mathbf{v}$. We then just write f as $(2\pi)^{-1} f_0(\rho)$. Furthermore, we impose the boundary condition

$$f_0(\rho = 0) = f_0(\rho = \infty) = 0,$$

since any zero value for the radius of curvature for two diverging trajectories will immediately grow with time and will never be reduced again to zero by collisions. The condition at infinite radius of curvature is required so that f will be properly integrable over all values of ρ. This equation is now easily solved by considering the regions $0 \le \rho \le a/2$ and $a/2 \le \rho \le \infty$ separately, and by requiring that f be continuous at $\rho = a/2$. Then one finds (see Exercise 18.2.)

$$f_0(\rho) = \begin{cases} 2na\exp(-2na\rho) & \text{for } \rho \ge a/2 \\ 2na\left[1 - \sqrt{1 - 4\rho^2/a^2}\right] & \text{for } 0 \le \rho \le a/2. \end{cases} \tag{18.21}$$

This expression can now be inserted in (18.10) to obtain

$$\lambda_+ = v\int_0^\infty d\rho \frac{1}{\rho} f_0(\rho)$$

$$= v\int_0^{a/2} d\rho \frac{1}{\rho}\left[1-\sqrt{1-\frac{4\rho^2}{a^2}}\right] + v\int_{a/2}^{\infty} d\rho \frac{1}{\rho}\exp(-2na\rho)$$

$$= 2na^2\left(\frac{v}{a}\right)\left[-\ln(2na^2) - \mathcal{C} + 1\right]. \qquad (18.22)$$

Here, $\mathcal{C} = 0.577\ldots$ is Euler's constant which occurs in the density expansion of the logarithmic integral function appearing on the right-hand side of (18.22). The functional form of the Lyapunov exponent, $\lambda_+ \sim -n \ln n$, was anticipated by Krylov who gave simple mean free path arguments for the exponential separation of trajectories, which led to this form.†

Finally, we mention that the Boltzmann equation method described here can be applied to computing the negative Lyapunov exponent for the two-dimensional Lorentz gas – by following the time-reversed motion. This method can also be extended to higher dimensions as well. Of special interest is the applications of the method to compute the Lyapunov exponents needed for transport coefficients discussed in Chapter 13. There one needs either Lyapunov exponents for the fractal repeller in order to use the escape-rate formalism, or the changes in the Lyapunov exponents produced by a thermostatted electric field to use the thermostat formalism. In both cases, the calculations can be done, with strikingly interesting results. We refer the reader to papers by van Beijeren *et al.*, listed below, for further details.

18.4 Semi-dispersing billiard systems

The Lorentz gas system we have been discussing is an example of a *dispersing* billiard system. By this term, we mean that if a small pencil of parallel trajectories impinges on a scatterer, the pencil will be expanded as a result of the collisions (see Fig. 18.4). This dispersing property is a characteristic of any convex scatterer. If the scatterers were flat plates instead of convex objects, and the collisions were specular, then the pencil of parallel trajectories would be reflected without expansion or contraction. A third possibility is that the scatterers are concave, in which case specular scattering would tend to focus the rays in the pencil, just as in

† These arguments are presented in the books by Krylov, edited by Sinai, and the book by Ma, mentioned below.

the optical properties of mirrors. Of course there are objects, such as circular cylinders, which have different properties in different directions. Here, a cross-section perpendicular to the long axis of the cylinder is convex while the direction along the long axis is flat. Such an object would be called a *semi-dispersing scatterer*, meaning that there are some directions in space in which an initial pencil is not expanded. This case is actually of major physical importance, as we now see.

Consider a gas of N hard spheres in d dimensions. If we were to imagine a configuration space of Nd dimensions, then this space has an interesting set of properties. We could describe the dynamics of the hard-sphere system in terms of the motion of one point in this hyperspace, with position $\mathbf{R} = (\mathbf{r}_1, \mathbf{r}_2, \ldots, \mathbf{r}_N)$ and velocity $\mathbf{V} = (\mathbf{v}_1, \mathbf{v}_2, \ldots, \mathbf{v}_N)$. The volume in configuration space which is excluded from the motion of the point at $(\mathbf{R}, \mathbf{V})$ is a collection of hypercylinders defined by the condition that $|\mathbf{r}_k - \mathbf{r}_l| \leq a$ for all pairs k, l, and a is the diameter of the spheres. Thus the 'hyper-point' moves freely in the configuration space until some pair of particles undergo a collision when $|\mathbf{r}_i - \mathbf{r}_j| = a$, for some pair of particles, i, j. A little imagination and reflection will convince the reader that although we are thinking of this process taking place in a high-dimensional configuration space, it is simply a collision of a point with a cylinder. Therefore, there will be some directions in this space which are dispersing and others which are not dispersing, and it is proper to think of the dynamics of a system of N hard spheres as that of a semi-dispersing billiard system. Of course, this general picture also holds if the particles interact with any short-range, purely repulsive potential.

Although this semi-dispersing property makes the dynamical theory for hard-sphere systems more complicated, it is not impossible to make progress in this direction. A number of ergodic theorems have been proved by Sinai and co-workers, and more recently by Simányi and Szász. Moreover, it is also possible to use kinetic theory and Boltzmann equation methods to compute various properties of the Lyapunov spectrum for hard-sphere and hard-disk systems. For example, van Zon and van Beijeren have been able to compute the largest Lyapunov exponent for a dilute gas of hard disks, and van Beijeren, Dorfman, and Latz have worked out methods to compute the sum of all the positive Lya-

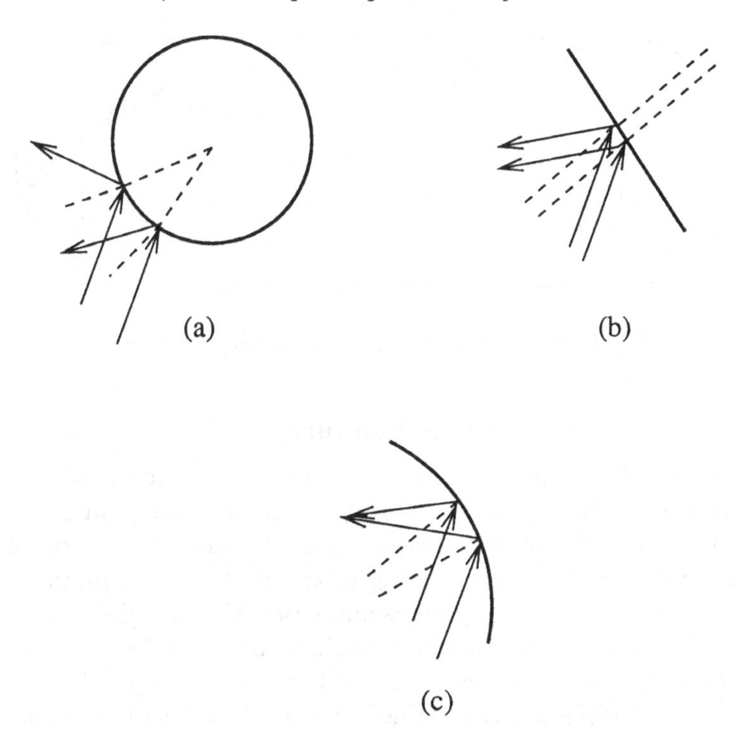

Fig. 18.4 Illustration of the reflection of parallel billiard trajectories from: (a) a convex surface; (b) a planar surface; and (c) a concave surface.

punov exponents for dilute gases of hard disks or hard spheres. The references are given below.

We mention that Bunimovich and Wojtkowski showed that it is possible to prove ergodic behavior for a particle moving in the interior of a two-dimensional space with partially concave boundaries, such as a stadium with flat sides and semi-circular caps at the ends (see Fig. 18.5). The motion is ergodic as long as there is some flat region on the sides of the stadium. The concave caps do indeed focus a pencil of trajectories but the rays go through the focus and start to expand again. This expansion leads to the ergodic behavior.

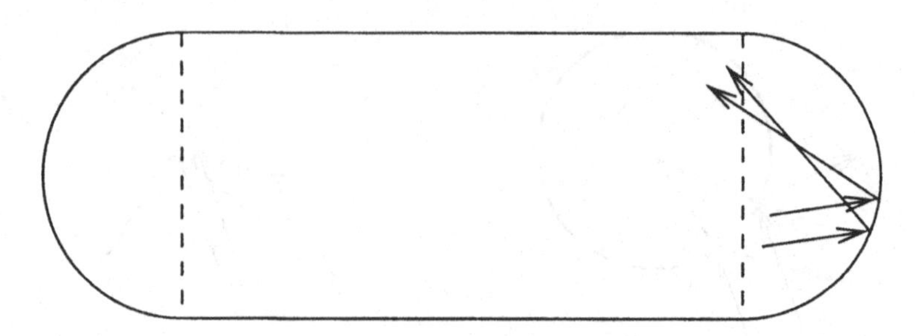

Fig. 18.5. Illustration of Bunimovich's stadium.

18.5 Summary

It is a remarkable fact that in this book we have come full circle! We started with the classical methods of Boltzmann in order to compute the transport properties of a dilute gas. Then, in order to arrive at a deeper understanding of the Boltzmann equation, we studied the roles of large systems and ensemble averaging, as in the Kac ring model, and the role of ergodic theory, as in the Boltzmann equation appearing in the theory of the baker's map. Ultimately, our understanding of irreversible behavior depends upon all of the concepts discussed here. Both the ergodicity of physical systems, and the laws of large numbers have a role to play in irreversible behavior and the precise mix of the features depends a great deal upon the system under discussion. Nevertheless, the main line of the argument has been that for systems with nice transport properties, such as hard-sphere or hard-disk systems, one must study their ergodic and mixing behavior in order to grasp the ideas that are at the heart of our understanding of the irreversible properties of mechanically reversible systems. It is therefore especially satisfactory that there is a deep consistency in the theory now.† Not only do we need the positivity of Lyapunov exponents to justify Boltzmann-like equations for irreversible behavior, but we can also use the Boltzmann equation to compute Lyapunov exponents and show that some of the exponents are indeed positive, as required. The further development of these ideas is likely to be a rich area of research in the near future.

† Although we still lack a proof of the ergodic behavior of a gas of identical billiard balls, and we lack a rigorous derivation of the extended Lorentz–Boltzmann equation.

18.6 Further reading

The fundamental theory for the chaotic behavior of dispersing billiards was provided by Sinai in his 1970 paper, reprinted in the collection *Dynamical Systems* [Sin91]. The discussion of the kinetic theory calculation of the Lyapunov exponents for a dilute, random Lorentz gas can be found in the papers of van Beijeren *et al.* [vBD95, vBD96, vBDC$^+$96, LvBD97, vBLD98]. Comparisons with numerical simulations are described in the papers of Dellago and Posch and co-workers [vBDC$^+$96, DP97]. A nice summary of Krylov's argument for logarithmic terms in the expression for the Lyapunov exponent for a dilute gas is given in the book by Ma [Ma85]. See also Krylov's book [Kry79]. A rigorous derivation of the Lorentz–Boltzmann equation in the low-density limit has been given by Boldrighini, Bunimovich, and Sinai [BBS83].

Discussions of the ergodic properties of semi-dispersing and of focusing billiards are given in the articles by Sinai *et al.*, and by Bunimovich in books edited by Sinai [Sin89, Sin91]. Wojtkowski has given a discussion of the stadium billiard and related topics [Woj85]. Further discussions of the ergodic properties of billiard-ball systems are to be found in the book by Katok and Strelcyn [KS87]. A very clear review of the interesting problems that occur when trying to establish ergodic theorems for billiard-ball systems has been given by Liverani and Wojtkowski [LW95]. Here one can find discussions of the difficulties caused by discontinuities in the scattering for the application of the Hopf method for proving ergodicity for such systems, as well as of the elegant and subtle methods used by Sinai and Chernov to overcome these difficulties [Sin91]. An interesting discussion of ergodicity in another discontinuous system, the sawtooth map, has been given by Vaienti [Vai92]. In the few cases of semi-dispersing billiard-ball systems where proofs are available, one can prove that the phase-space can be decomposed into a countable number of ergodic components, each of positive measure, and that the dynamics is of Bernoulli type on each component. More recent work has been carried out by the group of Kramli, Szász, and Simányi [Szá93]. Applications of kinetic theory methods to the calculation of the largest Lyapunov exponent of a dilute gas of hard disks has been given by van Zon, van Beijeren, and Dellago [vZvB98], and applications of kinetic theory to the calculation of the KS entropy of a dilute gas of hard disks or hard spheres has been given by van Beijeren *et al.* [vBDPD97, DLvB98].

18.7 Exercises

1. Derive the expression for the change in the radius of curvature due to collisions with a circular scatterer, (18.2).

2. Solve the equilibrium extended Lorentz–Boltzmann equation, (18.21), and obtain (18.22) for $na^2 \ll 1$.

19
What's next ?

We have now covered the background material needed to approach the literature on dynamical systems theory and nonequilibrium statistical mechanics. Here we list a few topics that you might find stimulating to think about. Some references are provided, but you should spend some time on the computer looking up relevant papers in areas that you find especially interesting.

Very nice overviews of this field of research are provided by D. Ruelle and Ya. G. Sinai in their paper 'From dynamical systems to statistical mechanics and back' [RS86]; in Ruelle's lecture notes, 'New theoretical results in nonequilibrium statistical mechanics' [Rue98]; and in the paper of G. Gallavotti, 'Chaotic dynamics, fluctuations, nonequilibrium ensembles' [Gal98].

A very beautiful and more advanced discussion of many of the topics covered in the later chapters of this book is provided in the monograph *Chaos, Scattering, and Statistical Mechanics* [Gas98], by P. Gaspard. Some general reviews of these subjects can also be found in papers by Gaspard; van Beijeren and Dorfman; Cohen; Dellago and Posch; Morriss, Dettmann and Rondoni, in *Physica* A, **240** Nos. 1–2 (1997), and in the *Chaos Focus Issue: Chaos and Irreversibility* [TGN98].

19.1 Billiard systems

In this book, we have only touched lightly the deep and rich subject of the dynamical theory of hard-sphere systems. This area has been developed by Sinai and co-workers and constitutes one of the most fascinating areas for study – it is a field of beautiful mathematics and of major physical interest. Some of the key papers are:

1. Several important papers by members of the Russian School, including that by Ya. G. Sinai, 'Dynamical systems with elastic reflections', and that by N. Chernov and Ya. G. Sinai, 'Ergodic properties of certain systems of two-dimensional discs and three-dimensional balls', are collected in the volume *Dynamical Systems, A Collection of Papers*, Ya. G. Sinai, ed. [Sin91];
2. G. Gallavotti and D. Ornstein, 'Billiards and Bernoulli schemes' [GO74];
3. D. Szász, 'Ergodicity of classical billiard balls' [Szá93];
4. S. Tabachnikov, *Billiards* [Tab95];
5. N. Simányi and D. Szász, 'Hard ball systems are fully ergodic' [SS96];
6. N. Simányi, 'Ergodicity of hard spheres in a box' [Sim97].

19.2 Model systems

There are an increasing number of studies of simple, reversible, deterministic systems like the baker's transformation where transport processes can be studied in great detail. Some surprising results have been obtained in connection with the hydrodynamic behavior of these systems. The following papers might be of some interest:

1. P. Gaspard, 'Diffusion, effusion, and chaotic scattering: an exactly solvable Liouvillian dynamics' [Gas92a];
2. P. Gaspard, 'What is the role of chaotic scattering in irreversible processes?' [Gas93];
3. R. Klages and J. R. Dorfman, 'Simple maps with fractal diffusion coefficients' [KD95];
4. S. Tasaki and P. Gaspard, 'Fick's law and fractality of nonequilibrium stationary states in a reversible multibaker map' [TG95];
5. P. Gaspard, 'Entropy production in open volume-preserving systems' [Gas97b].
6. J. Vollmer, T. Tél, and W. Breymann, 'Entropy balance in the presence of drift and diffusion currents: an elementary chaotic map approach' [VTB98];

19.3 Ruelle–Pollicott resonances

We have mentioned a number of times that there exist quantities called Ruelle–Pollicott resonances. These resonances appear as eigenvalues of the Frobenius–Perron operator, or, equivalently, as poles in the complex plane of the resolvent operator for the Frobenius–Perron equation. A study of these resonances leads to information about the transport properties of the system. For model systems, at least, one can say a great deal about these resonances. There is a detailed discussion of them in Gaspard's book [Gas98], and some interesting papers are to be found in:

1. D. Ruelle, 'Resonances of classical dynamical systems' [Rue86b];
2. D. Ruelle, 'Locating resonances for Axiom-A dynamical systems' [Rue86a];
3. P. Gaspard, 'r-adic one-dimensional maps and the Euler summation formula' [Gas92c];
4. P. Gaspard, 'Diffusion in uniformly hyperbolic one-dimensional maps and Appell polynomials' [Gas92b];
5. H. H. Hasegawa and W. C. Saphir, 'Decaying eigenstates for simple chaotic systems' [HS92a];
6. H. H. Hasegawa and W. C. Saphir, 'Nonequilibrium statistical mechanics of the baker map: Ruelle resonances and subdynamics' [HS92b];
7. H. H. Hasegawa and D. J. Driebe, 'Transport as a dynamical property of a simple map' [HD92];
8. H. H. Hasegawa and W. C. Saphir, 'Unitarity and irreversibility in chaotic systems' [HS92c];
9. M. Dörfle, 'Spectrum and eigenfunctions of the Frobenius–Perron operator of the tent map' [Dör85];
10. P. Gaspard, 'Hydrodynamic Modes as singular eigenstates of the Liouvillian dynamics: deterministic diffusion' [Gas96];
11. R. F. Fox, 'Unstable evolution of pointwise trajectory solutions to chaotic maps' [Fox95];
12. R. F. Fox, 'Construction of the Jordan basis for the baker map' [Fox97].

19.4 Thermostatted systems

Attempting to understand the dynamical behavior of thermostatted systems and the role of entropy production in them continues to be an area where considerable progress is being made. An important issue is the validity of the conjugate pairing rule, whereby the the sum of each pair of non-zero Lyapunov exponents may, in many circumstances, add up to the same value, independent of the index of the particular pair. For symplectic systems this sum is always zero, but for thermostatted systems, this sum is negative and depends on the value of the applied field. The conjugate pairing rule, when valid, allows for transport coefficients of many-particle systems to be determined by the values of just one pair of Lyapunov exponents, usually the most positive and the most negative ones. Some recent discussions have been provided in:

1. E. G. D. Cohen, 'Transport coefficients and Lyapunov exponents' [Coh95];
2. G. Gallavotti and E. G. D. Cohen, 'Dynamical ensembles and nonequilibrium statistical mechanics' [GC95a];
3. G. Gallavotti and E. G. D. Cohen, 'Dynamical ensembles in stationary states' [GC95b];
4. W. Breymann, T. Tél, and J. Vollmer, 'Entropy balance, time reversibility, and mass transport in dynamical systems' [BTV98];
5. G. P. Morriss and C. P. Dettmann, 'Thermostats: analysis and applications' [MD98];
6. T. Gilbert and J. R. Dorfman, 'Entropy production: from open volume preserving to dissipative systems' [GD98];
7. T. Gilbert, C. Ferguson, and J. R. Dorfman, 'Field driven thermostatted systems: a nonlinear, multi-baker map' [GFD99].

19.5 Other approaches to transport

The approaches to a fundamental theory of transport in fluids discussed in this book by no means exhaust all the possibilities. One paper, by N. I. Chernov and J. L. Lebowitz, 'Stationary nonequilibrium states in boundary driven Hamiltonian systems: shear flow' [CL97], describes a method by means of which a phase-space contraction in a system with a shear flow can be produced by a special kind of boundary condition. Another paper, by L. A.

Bunimovich and H. Spohn, 'Viscosity for a periodic two disk fluid: an existence proof' [BS96], takes a simulation method of Hoover and others and turns it into a nice, mathematically detailed, description of viscous flow. The model here is an infinite tiling of the plane by unit cells, each of which have two hard disks in them, the disks themselves cannot move from one cell to another, and the configurations and velocities in each cell are repeated in each cell of the tiling. This is an interesting and provocative modeling of a fluid flow, which, surprisingly, gives viscosities not too different from more realistic models.

19.6 Further applications of mathematical ideas

19.6.1 Differential geometry

In a very real sense, the natural mathematical language for classical mechanics is differential geometry and Riemannian manifolds. From this point of view, it is not surprising that an important early example of an ergodic system is the behavior of geodesics on surfaces of constant negative curvature. Recent literature in the applications of chaos to statistical mechanics has taken advantage of ideas in differential geometry to make interesting statements about physical systems. Some interesting papers on this subject are:

1. C. P. Dettmann and G. P. Morriss, 'Proof of conjugate pairing for an isokinetic thermostat' [DM96], as well as their paper on thermostats [MD98] mentioned above;
2. M. P. Wojtkowski and C. Liverani, 'Conformally symplectic dynamics and symmetry of the Lyapunov spectrum' [WL98];
3. P. Choquard, 'Variational principles for thermostatted systems' [Cho98];
4. L. Casetti, *Aspects of Dynamics, Geometry, and Statistical Mechanics in Hamiltonian Systems*, Ph. D. thesis [Cas97];
5. L. Casetti, C. Clementi, and M. Pettini, 'Riemannian theory of Hamiltonian chaos and Lyapunov exponents' [CCP96].
6. P. Cipriani and M. Di Bari 'Finsler geometric local indicator of chaos for single orbits in the Henon–Heiles Hamiltonian [CB98b];
7. P. Cipriani and M. Di Bari 'Dynamical instability and statistical behavior on N-body systems' [CB98a];

8. M. Di Bari and P. Cipriani ' Geometry and chaos on Riemannian and Finsler manifolds' [BC98].

19.6.2 Random matrix theory

The calculation of Lyapunov exponents for systems in more than one dimension often involves the calculation of the eigenvalues of products of random matrices. Consequently, this topic is of great interest to workers in the field, in addition to the interest in the theory of random matrices which characterizes much work in quantum chaos. Some useful references are:

1. A. Crisanti, G. Paladin, and A. Vulpiani, *Products of Random Matrices in Statistical Physics* [CPV93];
2. S. Sastry, 'Lyapunov spectra, instantaneous normal modes, and relaxation in the Lennard-Jones liquid' [Sas96];
3. A. Latz, H. van Beijeren, and J. R. Dorfman, 'Lyapunov spectrum and conjugate pairing rule for a thermostated Lorentz gas' [LvBD97].

19.7 Random perturbations

Here, we have taken the view that it is important to consider the dynamical behavior of an isolated system of particles. Of course, no laboratory system is truly isolated from its surroundings. Such systems are always subjected to stray fields, vibrations of the laboratory produced by people, other experiments, local construction projects, subways, etc. Computer simulations of molecular dynamics, too, are subject to random perturbations produced by the round-off errors in computer arithmetic.† It is therefore of interest to study the effects of random perturbations on dynamical systems and to see what features of our discussions are affected by these perturbations. One can imagine that under favorable circumstances, the random perturbations might enhance the mixing nature of a physical system. Next, one can imagine that random perturbations would destroy or 'smooth out' the fractal structures

† This is one reason for the interest in cellular automata lattice gases, which operate under a binary arithmetic and are not usually affected by round-off errors.

that seem to underly the transport and other properties of the system. Finally, one can imagine that small random perturbations could determine the phase-space measure, if there were more than one possibility, so that the most robust measure would be the physical one. It is important, therefore, to determine if the effects of random perturbations would be a serious complicating factor that would change the nature of our understanding of irreversible processes in fluids, or be an aid to the approach to equilibrium, making non-hyperbolic systems behave as if they were hyperbolic. Interesting discussions are to be found in the books:

1. M. I. Freidlin and A. D. Wentzell, *Random Perturbations of Dynamical Systems* [FW84];
2. Y. Kifer, *Random Perturbations of Dynamical Systems* [Kif88];
3. P.-D. Liu, *Smooth Ergodic Theory of Random Dynamical Systems* [Liu95];

and in the recent papers:

1. C. K. Cole and W. C. Schieve, 'Noise on the triadic Cantor set' [CS94];
2. G. Nicolis and D. Daems 'Nonequilibrium thermodynamics of dynamical systems' [ND96];
3. D. Ruelle, 'Positive entropy production in the presence of a random thermostat' [Rue97b];
4. D. Ruelle, 'Random smooth dynamical systems' [Rue96b];
5. G. Radons, 'Suppression of chaotic diffusion by quenched disorder' [Rad96];
6. G. Radons, 'Disorder phenomena in chaotic systems' [Rad98].

19.8 Quantum systems

We have hardly said a word about the quantum mechanical version of the theoretical developments presented here. This area is one in which a number of fundamental questions are still without answers, such as 'Is there such a thing as quantum chaos?'

The reader is advised to look at the papers collected in:

1. The special issue of the journal *Chaos*, Vol. 3, No. 4 (1993), devoted to chaotic scattering for a recent review of this subject;
2. G. Casati and B. Chirikov, eds., *Quantum Chaos* [CC95].

Important books are:

1. M. C, Gutzwiller, *Chaos in Classical and Quantum Mechanics* [Gut90];
2. M. L. Mehta, *Random Matrices* [Meh91];
3. F. Haake *Quantum Signatures of Chaos* [Haa91];
4. K. Nakamura, *Quantum Chaos* [Nak93].

Recent reviews of topics in quantum transport with close connections to quantum chaos are:

1. C. W. J. Beenakker, 'Random matrix theory of transport' [Bee97];
2. T. Guhr, A. Müller-Groeling, and H. A. Weidenmüller, 'Random-matrix theories in quantum physics: common concepts' [GMGW98].

19.9 Experimental studies

Gaspard has proposed a method by which microscopic chaos, as opposed to macroscopic chaos, can be detected in a laboratory experiment. The main idea is to follow a long trajectory of a Brownian particle in solution and then use a variety of techniques to infer an effective KS-entropy from the data. An experiment has been carried out and a positive KS entropy has been obtained as a result of the analysis. This result is certainly consistent with the idea that the Brownian particle is undergoing chaotic motion. No doubt further, and more sensitive, experiments will be carried out in the future to provide evidence for the chaoticity of microscopic motions in a fluid. The references are:

1. P. Gaspard, 'Can we observe microscopic chaos in the laboratory?' [Gas97a]; and
2. P. Gaspard, M. E. Briggs, M. K. Francis, J. V. Sengers, R.W. Gammon, J. R. Dorfman, and R. V. Calabrese, 'Experimental evidence for microscopic chaos' [GBF$^+$98]. See also the companion paper, with a discussion of this experiment, in the same issue of *Nature*, 'Brownian motion and microscopic chaos' by Dürr and Spohn [DS98].

The Gallavotti–Cohen fluctuation theorem has been, and will certainly continue to be, subject to experimental as well as numerical verification. A recent paper which looked for evidence of the theorem by measuring local temperature fluctuations in a Rayleigh–Bénard convection cell is:

1. S. Ciliberto and C. Laroche, 'An experimental test of the Gallavotti–Cohen fluctuation theorem' [CL98].

19.10 Problems for the future

The purpose of this book is to provide the reader with an introduction to the basic ideas and techniques in dynamical systems theory that are useful for a deeper understanding of nonequilibrium statistical mechanics. It is in the nature of textbooks to present the material in an organized and coherent way, in so far as possible, so that the reader obtains a broad view of the subject and the major results obtained. However, in a young subject such as this, there are many problems to be resolved that may provide some challenges to the reader as well as to the author. Here are just a few:

1. Van Beijeren and Cohen have pointed out that the wind-tree model is an anomaly from the point of view of the ideas presented here.† As we saw in Exercise 2.3, it is possible to compute the diffusion coefficient for the moving particle in the wind-tree model, when the trees are placed at random in the plane at low density and without overlapping. The anomalous nature of this model resides in the fact that although there is a well-defined transport coefficient, there are no positive Lyapunov exponents associated with the moving particle, so that the motion of the particle is not chaotic in the usual sense. This is a consequence of the non-dispersing nature of the collisions of the moving particle with the trees, since the sides of the trees are flat. Therefore, our idea of associating transport with chaotic motion is not complete and there are cases, such as the wind-tree model, where this picture appears to break down. In this case one might suspect that the random distribution of trees is important.
2. We have made a point of associating the approach of distribution functions to their equilibrium values with the stretching

† I thank H. van Beijeren and E. G. D. Cohen for communicating this interesting and important observation to me.

of phase-space regions along unstable directions, and then projecting phase-space distributions on the subspace of macroscopic variables. However, we have only described this process in any detail for simple two-dimensional models, such as the baker's map, and the Arnold cat map. One should remain a bit sceptical about this picture until it is verified for macroscopic systems with many degrees of freedom, such as one would study in the laboratory, or in a computer simulation. In the latter case, the number of degrees of freedom may not be so very large, but certainly greater than two.

3. We have said very little about the quantum versions of the classical chaotic dynamics discussed here. However, nature is fundamentally quantum mechanical. So in so far as this material has any relevance for natural processes, a quantum version is certainly necessary. One would very much like to know something about the quantum mechanical quantities that play a role similar to that of Lyapunov exponents, KS entropies, etc., in classical physics. This is all open for investigation.

Bibliography

[AA68] V. I. Arnold and A. Avez. *Ergodic Problems of Classical Mechanics.* W. A. Benjamin, Inc., New York, 1968.

[AAC90a] R. Artuso, E. Aurell, and P. Cvitanović. Recycling of strange sets, I: Cycle expansions. *Nonlinearity*, **3**:325, 1990.

[AAC90b] R. Artuso, E. Aurell, and P. Cvitanović. Recycling of strange sets, II: Applications. *Nonlinearity*, **3**:361, 1990.

[AE97] C. Appert and M. H. Ernst. Chaos properties and localization in Lorentz lattice gases. *Physical Review E*, **56**:5106, 1997.

[Ano67] D. V. Anosov. Geodesic flows on closed Riemannian manifolds of negative curvature. In *Proceedings of the Steklov Institute of Mathematics*, volume 90, Providence, 1967.

[Arn89] V. I. Arnold. *Mathematical Methods of Classical Mechanics,* 2nd edition. Springer-Verlag, New York, 1989.

[AS67] D. V. Anosov and Ya. G. Sinai. Certain smooth ergodic systems. *Uspekhi Matematicheskikh Nauk, (Russian Mathematical Surveys)*, **22**:107, 1967.

[ASY97] K. T. Alligood, T. D. Sauer, and J. A. Yorke. *Chaos, An Introduction to Dynamical Systems.* Springer-Verlag, New York, 1997.

[AvBED96] C. Appert, H. van Beijeren, M. H. Ernst, and J. R. Dorfman. Thermodynamic formalism in the thermodynamic limit: diffusive systems with static disorder. *Physical Review E*, **54**:R1013, 1996.

[AvBED97] C. Appert, H. van Beijeren, M. H. Ernst, and J. R. Dorfman. Thermodynamic formalism and localization in Lorentz gases and hopping models. *Journal of Statistical Physics*, **87**:1253, 1997.

[Bal97] R. Balescu. *Statistical Dynamics: Matter out of Equilibrium.* Imperial College Press, London, 1997.

[Bal99] V. Baladi. The magnet and the butterfly: thermodynamic formalism and the ergodic theory of chaotic dynamics. In J-P. Pier, editor, *Développement des Mathématiques au Cours de la Seconde Moitié du XXe Siècle*, Basel, 1999. Birkhäuser Publishing Co.

[BBS83] C. Boldrighini, L. A. Bunimovich, and Ya. G. Sinai. On the Boltzmann equation for the Lorentz gas. *Journal of Statistical Physics*, **32**:477, 1983.

[BC98] M. Di Bari and P. Cipriani. Geometry and chaos on Riemann and Finsler manifolds. *Planetary and Space Science*, **46**:1543, 1998.

[Bed91] T. Bedford. Applications of dynamical systems theory to fractals – a study of cookie-cutter Cantor sets. In J. Bélair and S. Dubuc, editors, *Fractal Geometry and Analysis*, Dordrecht, 1991. Kluwer Academic Publishers.

[Bee97] C. W. J. Beenakker. Random matrix theory of quantum transport. *Reviews of Modern Physics*, **69**:731, 1997.

[Ber77] B. Berne, editor. *Statistical Mechanics, Part B.* Plenum Publishing Co., New York, 1977.

[Ber87] M. Berry. Regular and irregular motion. In R. S. MacKay and J. D. Meiss, editors, *Hamiltonian Dynamical Systems.* Adam Hilger, Bristol, England, 1987.

[BGG97] F. Bonetto, G. Gallavotti, and P. L. Garrido. Chaotic principle: An experimental test. *Physica D*, **105**:226, 1997.

[Bog62] N. N. Bogoliubov. Problems of a dynamical theory in statistical physics. In *Studies in Statistical Mechanics*, volume 1, page 5. North Holland Publishing Co., Amsterdam, 1962.

[Bol95] L. Boltzmann. *Lectures on Gas Theory.* Dover Publishing Co., New York, 1995. Translation by S. Brush.

[Bow75] R. Bowen. *Equilibrium States and the Ergodic Theory of Anosov Diffeomorphisms*, volume **470** of *Lecture Notes in Mathematics.* Springer-Verlag, Berlin, 1975.

[Bow78a] R. Bowen. Markov partitions are not smooth. *Proceedings of the American Mathematical Society*, **71**:130, 1978.

[Bow78b] R. Bowen. *On Axiom-A Diffeomorphisms*, volume **35**. American Mathematical Society, Providence, 1978.

[BPR97] F. Bagnoli, P. Palmerini, and R. Rechtman. Algorithmic mapping from criticality to self-organized criticality. *Physical Review E*, **55**:3970, 1997.

[BR75] R. Bowen and D. Ruelle. The ergodic theory of Axiom-A flows. *Inventiones Mathematicae*, **29**:181, 1975.

[BR87] T. Bohr and D. Rand. The entropy function for characteristic exponents. *Physica D*, **25**:387, 1987.

[Bri95] J. Bricmont. Science of chaos or chaos in science? *Physicalia Magazine*, **17**:159, 1995.

[Bru72] S. Brush. *Kinetic Theory, Vols. 1, 2, 3.* Pergamon Press, New York, 1972.

[Bru76] S. Brush. *The Kind of Motion We Call Heat.* North-Holland

Publishing Co., Amsterdam, 1976.

[BS93] C. Beck and F. Schlögl. *Thermodynamic Formalism.* Cambridge University Press, Cambridge, 1993.

[BS96] L. A. Bunimovich and H. Spohn. Viscosity for a periodic two disk fluid: an existence proof. *Communications in Mathematical Physics*, **176**:661, 1996.

[BS99] R. K. Bowles and R. J. Speedy. Five disks in a box. *Physica A*, **262**:76, 1999.

[BTV96] W. Breymann, T. Tél, and J. Vollmer. Entropy production for open dynamical systems. *Physical Review Letters*, **77**:2945, 1996.

[BTV98] W. Breymann, T. Tél, and J. Vollmer. Entropy balance, time reversibility, and mass transport in dynamical systems. *Chaos*, **8**:396, 1998.

[BY91] J.-P. Boon and S. Yip. *Molecular Hydrodynamics.* Dover Publishing Co., New York, 1991.

[C+97] P. Cvitanović *et al. Classical and Quantum Chaos: A Cyclist Treatise.* Neils Bohr Institute, Copenhagen, 1997.

[Cas97] L. Casetti. *Aspects of Dynamics, Geometry and Statistical Mechanics in Hamiltonian Systems.* Ph. D. thesis, Scuola Normale Superiore, Pisa, 1997.

[Caw91] E. Cawley. Smooth Markov partitions and toral automorphisms. *Ergodic Theory and Dynamical Systems*, **11**:633, 1991.

[CB98a] P. Cipriani and M. Di Bari. Dynamical instability and statistical behavion of N-body systems. *Plametary and Space Science*, **46**:1499, 1998.

[CB98b] P. Cipriani and M. Di Bari. Finsler geometric local indicator of chaos for single orbits in the Henon–Heiles Hamiltonian. *Physical Review Letters*, **81**:5532, 1998.

[CC70] S. Chapman and T. G. Cowling. *The Mathematical Theory of Non-Uniform Gases.* Cambridge University Press, 1970.

[CC95] G. Casati and B. Chirikov, editors. *Quantum Chaos.* Cambridge University Press, Cambridge, 1995.

[CCP96] L. Casetti, C. Clementi, and M. Pettini. Riemannian theory of Hamiltonian chaos and Lyapunov exponents. *Physical Review E*, **54**:5969, 1996.

[CE80] P. Collet and J.-P. Eckmann. *Iterated Maps on the Interval as Dynamical Systems.* Birkhäuser, Boston, 1980.

[CE91] P. Cvitanović and B. Eckhardt. Periodic orbit expansions for classical smooth flows. *Journal of Physics A*, **24**:L237, 1991.

[CELS93] N. I. Chernov, G. L. Eyink, J. L. Lebowitz, and Ya. G. Sinai. Steady state electric conductivity in the periodic Lorentz gas. *Communications in Mathematical Physics*, **154**:569, 1993.

[CGS92] P. Cvitanović, P. Gaspard, and T. Schreiber. Investigation of the Lorentz gas in terms of periodic orbits. *Chaos*, **2**:85, 1992.

[CH93] M. C. Cross and P. C. Hohenberg. Pattern formation outside of equilibrium. *Reviews of Modern Physics*, **65**:851, 1993.

[Cho98] P. Choquard. Variational principles for thermostatted systems. *Chaos*, **8**:350, 1998.

[CL97] N. I. Chernov and J. L. Lebowitz. Stationary non-equilibrium states in boundary driven Hamiltonian systems: shear flow. *Journal of Statistical Physics*, **86**:953, 1997.

[CL98] S. Ciliberto and C. Laroche. An experimental test of the Gallavotti–Cohen fluctuation theorem. In A. Vulpiani, M. Serva, G. Parisi, L.Peliti, and L. Pietronero, editors, *Disorder and Chaos, Proceedings of the Conference in Honor of Giovanni Paladin*, Les Editions de Physique, IV, 1998.

[CM97a] N. Chernov and R. Markarian. Anosov maps with rectangular holes. *Boletim de Sociedade Brasileira de Matmatica*, **28**:315, 1997.

[CM97b] N. Chernov and R. Markarian. Ergodic properties of Anosov maps with rectangular holes. *Boletim de Sociedade Brasileira de Matmatica*, **28**:271, 1997.

[Coh93] E. G. D. Cohen. Fifty years of kinetic theory. *Physica A*, **194**:229, 1993.

[Coh95] E. G. D. Cohen. Transport coefficients and Lyapunov exponents. *Physica A*, **213**:293, 1995.

[Coh97a] E. G. D. Cohen. Bogoliubov and kinetic theory: the Bogoliubov equations. *Mathematical Models and Methods in the Exact Sciences*, **7**:909, 1997.

[Coh97b] E. G. D. Cohen. Boltzmann and statistical mechanics. *Atti dei Convegni Lincei*, **131**:11, 1997.

[Coh97c] E. G. D. Cohen. Dynamical ensembles in statistical mechanics. *Physica A*, **240**:43, 1997.

[CPV93] A. Crisanti, G. Paladin, and A. Vulpiani. *Products of Random Matrices in Statistical Physics.* Springer-Verlag, New York, 1993.

[CS94] C. K. Cole and W. C. Schieve. Noise on the triadic Cantor set. *Physica D*, **70**:302, 1994.

[CvB*et al.*nt] E. G. D. Cohen and H. van Beijeren *et al.*, editors. *Fundamental Problems in Statistical Mechanics.* North Holland Publishing Co., Amsterdam, 1962–present.

[DEJ95] J. R. Dorfman, M. H. Ernst, and D. Jacobs. Dynamical chaos in the Lorentz lattice gas. *Journal of Statistical Physics*, **81**:497, 1995.

[DG95] J. R. Dorfman and P. Gaspard. Chaotic scattering theory of transport and reaction-rate coefficients. *Physical Review E*, **51**:28, 1995.

[DKS94] J. R. Dorfman, T. R. Kirkpatrick, and J. V. Sengers. Generic long range correlations in fluids. In H. L. Strauss, G. T. Babcock, and S. R. Leone, editors, *Annual Review of Physical Chemistry*, volume 45, page 213. Annual Reviews, Inc., Palo Alto, 1994.

[DLvB98] J. R. Dorfman, A. Latz, and H. van Beijeren. BBGKY hierarchy methods for sums of Lyapunov exponents for dilute gases. *Chaos*, **8**:444, 1998.

[DM96] C. P. Dettmann and G. P. Morriss. Proof of conjugate pairing for an isokinetic thermostat. *Physical Review E*, **53**:R5545, 1996.

[Dor81] J. R. Dorfman. Some recent developments in the kinetic theory of gases. In H. J. Raveché, editor, *Perspectives in Statistical Physics*, page 23. North Holland Publishing Co., Amsterdam, 1981.

[Dör85] M. Dörfle. Spectrum and eigenfunctions of the Frobenius–Perron operator of the tent map. *Journal of Statistical Physics*, **40**:93, 1985.

[Dor98] J. R. Dorfman. Deterministic chaos and the foundations of kinetic theory. *Physics Reports*, **301**:151, 1998.

[DP91] A. DeMasi and E. Presutti. *Mathematical Methods for Hydrodynamic Limits. Lecture Notes in Mathematics*, No. 1501. Springer-Verlag, Berlin, 1991.

[DP97] Ch. Dellago and H. A. Posch. Lyapunov spectrum and the conjugate pairing rule for a thermostated random Lorentz gas: numerical simulations. *Physical Review Letters*, **78**:211, 1997.

[DS98] D. Dürr and H. Spohn. Brownian motion and microscopic chaos. *Nature*, **394**:831, 1998.

[DvB77] J. R. Dorfman and H. van Beijeren. The kinetic theory of gases. In B. Berne, editor, *Statistical Mechanics, Part B*, page 65. Plenum Press, New York, 1977.

[DvB97] J. R. Dorfman and H. van Beijeren. Dynamical systems theory and transport coefficients: a survey with applications to Lorentz gases. *Physica A*, **240**:12, 1997.

[ECM90] D. J. Evans, E. G. D. Cohen, and G. P. Morriss. Viscosity of a simple fluid from its maximal Lyapunov exponents. *Physical Review A*, **42**:5990, 1990.

[ECM93] D. J. Evans, E. G. D. Cohen, and G. P. Morriss. Probability of second law violations in steady shearing flows. *Physical Review Letters*, **71**:2401, 1993.

[EDNJ95] M. H. Ernst, J. R. Dorfman, R. Nix, and D. Jacobs. Mean-field theory for Lyapunov exponents and Kolmogorov–Sinai entropy in Lorentz lattice gases. *Physical Review Letters*, **74**:4416,

1995.

[EE59] P. Ehrenfest and T. Ehrenfest. *The Conceptual Foundations of the Statistical Approach in Mechanics.* Cornell University Press, Ithaca, 1959.

[EM90] D. J. Evans and G. P. Morriss. *Statistical Mechanics of Non-equilibrium Liquids.* Academic Press, London, 1990.

[Ern91] M. H. Ernst. Statistical mechanics of cellular automata fluids. In J. P. Hansen, D. Levesque, and J. Zinn-Justin, editors, *Liquids, Freezing and Glass Transition*, page 45. Elsevier Science Publishers, Dordrecht, 1991.

[Ern98] M. H. Ernst. Bogoliubov–Choh–Uhlenbeck theory: Cradle of modern kinetic theory. In W. Sung, I. Chang, B. Kahng, C. Kim, and J-H. Oh, editors, *Progress in Statistical Physics*, page 3. World Scientific Publishing Co., Singapore, 1998.

[EvV89a] M. H. Ernst and G. A. van Velzen. Long-time tails in Lorentz lattice gases. *Journal of Statistical Physics*, **57**:455, 1989.

[EvV89b] M. H. Ernst and G. A. van Velzen. Lorentz lattice gas. *Journal of Physics A: Math. Gen.*, **22**:4611, 1989.

[EvVB89] M. H. Ernst, G. A. van Velzen, and P. Binder. Breakdown of the Boltzmann equation in cellular automata lattice gases. *Physical Review A*, **39**:4327, 1989.

[Fal93] K. Falconer. *Fractal Geometry: Mathematical Foundations and Applications.* John Wiley, New York, 1993.

[Fal97] K. Falconer. *Techniques in Fractal Geometry.* John Wiley and Sons, New York, 1997.

[Fey67] R. Feynman. *The Character of Physical Law.* MIT Press, Cambridge, MA., 1967.

[FK72] J. H. Ferziger and H. G. Kaper. *Mathematical Theory of Transport Processes in Gases.* North Holland Publishing Co., Amsterdam, 1972.

[Fox95] R. F. Fox. Unstable evolution of pointwise trajectory solutions to chaotic maps. *Chaos*, **5**:619, 1995.

[Fox97] R. F. Fox. Construction of the Jordan basis for the baker map. *Chaos*, **7**:254, 1997.

[FW84] M. I. Freidlin and A. D. Wentzell. *Random Perturbations of Dynamical Systems.* Springer-Verlag, New York, 1984.

[Gal95] G. Gallavotti. Ergodicity, ensembles, irreversibility in Boltzmann and beyond. *Journal of Statistical Physics*, **78**:1571, 1995.

[Gal96] G. Gallavotti. Extension of Onsager's reciprocity to large fields and the chaotic hypothesis. *Physical Review Letters*, **77**:4334, 1996.

[Gal98] G. Gallavotti. Chaotic dynamics, fluctuations, nonequilibrium ensembles. *Chaos*, **8**:384, 1998.

[Gas92a] P. Gaspard. Diffusion, effusion, and chaotic scattering: An exactly solvable Liouvillian dynamics. *Journal of Statistical Physics*, **68**:673, 1992.

[Gas92b] P. Gaspard. Diffusion in uniformly hyperbolic one-dimensional maps and Appell polynomials. *Physics Letters A*, **168**:13, 1992.

[Gas92c] P. Gaspard. r-adic one-dimensional maps and the Euler summation formula. *Journal of Physics A: Math. Gen*, **25**:L483, 1992.

[Gas93] P. Gaspard. What is the role of chaotic scattering in irreversible processes? *Chaos*, **3**:427, 1993.

[Gas95] P. Gaspard. $\hbar$-expansion for quantum trace formulas. In G. Casati and B. Chirikov, editors, *Quantum Chaos*, page 385. Cambridge University Press, Cambridge, 1995.

[Gas96] P. Gaspard. Hydrodynamic modes as singular eigenstates of the Liouvillian dynamics: deterministic diffusion. *Physical Review E*, **53**:4379, 1996.

[Gas97a] P. Gaspard. Can we observe microscopic chaos in the laboratory? *Advances in Chemical Physics*, **XCIX**:369, 1997.

[Gas97b] P. Gaspard. Entropy production in open, volume preserving systems. *Journal of Statistical Physics*, **89**:1215, 1997.

[Gas98] P. Gaspard. *Chaos, Scattering, and Statistical Mechanics*. Cambridge University Press, Cambridge, 1998.

[GB92] P. Gaspard and F. Baras. Dynamical chaos underlying diffusion in the classical Lorentz gas. In M. Maréschal and B. L. Holian, editors, *Microscopic Simulations of Complex Hydrodynamic Phenomena*, volume 292 of *NATO ASI Series*, page 301, New York, 1992.

[GB95] P. Gaspard and F. Baras. Chaotic scattering and diffusion in the Lorentz gas. *Physical Review E*, **51**:5332, 1995.

[GBF$^+$98] P. Gaspard, M. E. Briggs, M. K. Francis, J. V. Sengers, R. W. Gammon, J. R. Dorfman, and R. V. Calabrese. Experimental evidence for microscopic chaos. *Nature*, **394**:865, 1998.

[GC95a] G. Gallavotti and E. G. D. Cohen. Dynamical ensembles in non-equilibrium statistical mechanics. *Physical Review Letters*, **74**:2694, 1995.

[GC95b] G. Gallavotti and E. G. D. Cohen. Dynamical ensembles in stationary states. *Journal of Statistical Physics*, **80**:931, 1995.

[GD95] P. Gaspard and J. R. Dorfman. Chaotic scattering theory, thermodynamic formalism and transport coefficients. *Physical Review E*, **52**:3525, 1995.

[GD98] T. Gilbert and J. R. Dorfman. Entropy production: from open volume preserving to dissipative systems. *submitted to Journal of Statistical Physics*, 1998. archived at chao-dyn@xyz.lanl.gov, No. 9804009.

[GFD99] T. Gilbert, C. Ferguson, and J. R. Dorfman. Field driven thermostatted systems: A non-linear multi-baker map. *Physical Review E*, **59**, 1999.

[GLS98] S. Goldstein, J. L. Lebowitz, and Ya. Sinai. Remark on the (non)convergence of ensemble densities in dynamical systems. *Chaos*, **8**:393, 1998.

[GMGW98] T. Guhr, A. Müller-Groeling, and H. A. Weidenmüller. Random-matrix theories in quantum physics: common concepts. *Physics Reports*, **299**:189, 1998.

[GN90] P. Gaspard and G. Nicolis. Transport properties, Lyapunov exponents, and entropy per unit time. *Physical Review Letters*, **65**:1693, 1990.

[GO74] G. Gallavotti and D. Ornstein. Billiards and Bernoulli schemes. *Communications in Mathematical Physics*, **38**:83, 1974.

[GOY88] C. Grebogi, E. Ott, and J. A. Yorke. Unstable periodic orbits and the dimension of chaotic attractors. *Physical Review A*, **37**:1711, 1988.

[GR97] G. Gallavotti and D. Ruelle. SRB states and non-equilibrium statistical mechanics close to equilibrium. *Communications in Mathematical Physics*, **190**:279, 1997.

[Gut90] M. C. Gutzwiller. *Chaos in Classical and Quantum Mechanics.* Springer-Verlag, New York, 1990.

[GW93] P. Gaspard and X.-J. Wang. Noise, chaos, and (ϵ, τ)-entropy per unit time. *Physics Reports*, **235**:291, 1993.

[Haa91] F. Haake. *Quantum Signatures of Chaos.* Springer-Verlag, Berlin, 1991.

[Hal56] P. R. Halmos. *Ergodic Theory.* Chelsea Publishing Co., New York, 1956.

[HD92] H. H. Hasegawa and D. J. Driebe. Transport as a dynamical property of a simple map. *Physics Letters A*, **168**:18, 1992.

[HH77] P. C. Hohenberg and B. I. Halperin. Theory of dynamic critical phenomena. *Reviews of Modern Physics*, **49**:435, 1977.

[Hoo91] W. G. Hoover. *Computational Statistical Mechanics.* Elsevier Science Publishers, Amsterdam, 1991.

[HS92a] H. H. Hasegawa and W. C. Saphir. Decaying eigenstates for simple chaotic systems. *Physics Letters A*, **161**:471, 1992.

[HS92b] H. H. Hasegawa and W. C. Saphir. Non-equilibrium statistical mechanics of the baker map: Ruelle resonances and subdynamics. *Physics Letters A*, **161**:477, 1992.

[HS92c] H. H. Hasegawa and W. C. Saphir. Unitarity and irreversibility in chaotic systems. *Physical Review A*, **46**:7401, 1992.

[HY93] B. R. Hunt and J. A. Yorke. Maxwell on chaos. *Nonlinear Science Today*, **3**:1, 1993.

[Jac90] E. A. Jackson. *Perspectives of Nonlinear Dynamics*, volume 2. Cambridge University Press, Cambridge, 1990.

[Kac76] M. Kac. *Probability and Related Topics in Physical Science.* American Mathematical Society, New York, 1976.

[Kat81] A. Katok. Dynamical systems with hyperbolic structure. In *Three Papers on Dynamical Systems, A. G. Kusnirenko, A. Katok, and V. M. Alekseev.* American Mathematical Society, 1981.

[Kat82] A. Katok. Entropy and closed geodesics. *Ergodic Theory and Dynamical Systems*, **2**:339, 1982.

[KD95] R. Klages and J. R. Dorfman. Simple maps with fractal diffusion coefficients. *Physical Review Letters*, **74**:387, 1995.

[Kel98] G. Keller. *Equilibrium States in Ergodic Theory.* Number 42 in London Mathematical Society Student Texts. Cambridge University Press, Cambridge, 1998.

[KG85] H. Kantz and P. Grassberger. Repellers, semi-attractors, and long-lived chaotic transients. *Physica D*, **17**:75, 1985.

[KH95] A. Katok and B. Hasselblatt. *Introduction to the Modern Theory of Dynamical Systems.* Cambridge University Press, Cambridge, 1995.

[Kif88] Y. Kifer. *Random Perturbations of Dynamical Systems.* Birkhäuser Publishing Co., Boston, 1988.

[KKPW90] A. Katok, G. Kneiper, M. Pollicott, and H. Weiss. Differentiability of entropy for Anosov and geodesic flows. *Bulletin of the American Mathematical Society*, **22**:285, 1990.

[Kla96] R. Klages. *Deterministic Diffusion in One-dimensional Chaotic Dynamical Systems.* Ph. D. thesis, Wissenschaft and Technik Verlag, Berlin, Germany, 1996.

[Kry79] N. S. Krylov. *Works on the Foundations of Statistical Mechanics.* Princeton University Press, Princeton, 1979.

[KS87] A. Katok and J. Strelcyn. *Smooth Maps with Singularities: Invariant Manifolds, Entropy, and billiards.* Number 1222 in Lecture Notes in Mathematics. Springer-Verlag, New York, 1987.

[KT84] L. P. Kadanoff and C. Tang. Escape from strange repellers. *Proceedings of the National Academy of Sciences, USA*, **81**:1276, 1984.

[KTH91] R. Kubo, M. Toda, and N. Hashitsume. *Statistical Physics*, 2nd edition, volume II. Springer-Verlag, New York, 1991.

[Kur98] J. Kurchan. Fluctuation theorem for stochastic dynamics. *Journal of Physics A: Math. Gen.*, **31**:3719, 1998.

[Lan76] O.E. Lanford. On a derivation of the Boltzmann equation. *Astérisque*, **40**:117, 1976.

[Lan81] O. E. Lanford. Hard-sphere gas in the Boltzmann–Grad limit. *Physica A*, **106**:70, 1981.

[Lan83] O. E. Lanford. Introduction to the mathematical theory of dynamical systems. In G. Iooss, R. H. G. Hellemann, and R. Stora, editors, *Chaotic Behavior of Deterministic Systems*, page 4. North Holland Publishing Co., Amsterdam, 1983.

[Leb93] J. L. Lebowitz. Macroscopic laws, microscopic dynamics, time's arrow, and Boltzmann's entropy. *Physica A*, **194**:1, 1993.

[Liu95] P.-D. Liu. *Smooth Ergodic Theory of Random Dynamical Systems*. Lecture Notes in Mathematics. Springer-Verlag, New York, 1995.

[Liv95] C. Liverani. Decay of correlations. *Annals of Mathematics*, **142**:239, 1995.

[LM94] A. Lasota and M. C. Mackey. *Chaos, Fractals, and Noise.* Springer-Verlag, New York, 1994.

[LP67] J. L. Lebowitz and J. K Percus. Kinetic equations and density expansions: exactly solvable one-dimensional system. *Physical Review*, **155**:122, 1967.

[LPS68] J. L. Lebowitz, J. K. Percus, and J. Sykes. Time evolution of the total distribution function of a one-dimensional system of hard rods. *Physical Review*, **171**:224, 1968.

[LS98] J. L. Lebowitz and H. Spohn. A Gallavotti–Cohen type symmetry in the large deviation functional for stochastic dynamics. *To be published*, 1998.

[LV93] D. Levesque and L. Verlet. Molecular dynamics and time reversibility. *Journal of Statistical Physics*, **72**:519, 1993.

[LvBD97] A. Latz, H. van Beijeren, and J. R. Dorfman. Lyapunov spectrum and conjugate pairing rule for a thermostated random Lorentz gas: kinetic theory. *Physical Review Letters*, **78**:207, 1997.

[LW95] C. Liverani and M. Wojtkowski. Ergodicity in Hamiltonian systems. In *Dynamics Reported (New Series)*, volume 4, page 130. Springer-Verlag, New York, 1995.

[LY85] F. Ledrappier and L-S. Young. The metric entropy of diffeomorphisms. *Annals of Mathematics*, **122**:509, 1985.

[Ma85] S. K. Ma. *Statistical Mechanics.* World Scientific, Singapore, 1985.

[Mac92] M. C. Mackey. *Time's Arrow: The Origin of Thermodynamic Behavior.* Springer-Verlag, New York, 1992.

[Mae98] C. Maes. The fluctuation theorem as a Gibbs property. *To be published*, 1998. archived at math-ph@xxx.lanl.gov, No. 9812015.

[Mar77] B. Marcus. Ergodic properties of horocycle flows for surfaces of constant negative curvature. *Annals of Mathematics*, **105**:81, 1977.

[Mar78] B. Marcus. The horocycle flow is mixing of all degrees. *Inventiones Mathematica*, **46**:201, 1978.

[Mar97] M. Maréschal, editor. *The Microscopic Approach to Complexity in Nonequilibrium Molecular Simulations*, volume **240**, *Physica* A, 1997.

[MB77] B. Marcus and R. Bowen. Unique ergodicity for horocycle foliations. *Israel Journal of Mathematics*, **26**:43, 1977.

[McL89] J. A. McLennan. *Introduction to Non-Equilibrium Statistical Mechanics*. Prentice Hall, Englewood Cliffs, New Jersey, 1989.

[MD98] G. P. Morriss and C. P Dettmann. Thermostats: analysis and application. *Chaos*, **8**:321, 1998.

[MDR97] G. P. Morriss, C. Dettmann, and L. Rondoni. Recent results for the thermostated Lorentz gas. *Physica A*, **240**:84, 1997.

[MECvB89] G. P. Morriss, D. J. Evans, E. G. D. Cohen, and H. van Beijeren. Linear response of phase space trajectories to shearing. *Physical Review Letters*, **62**:1579, 1989.

[Meh91] M. L. Mehta. *Random Matrices*. Academic Press, New York, 1991.

[MM87] R. S. MacKay and J. D. Meiss, editors. *Hamiltonian Dynamical Systems*. Adam Hilger, Bristol, England, 1987.

[MR96] G. P. Morriss and L. Rondoni. Equivalence of non-equilibrium ensembles. *Physica A*, **233**:767, 1996.

[Nak93] K. Nakamura. *Quantum Chaos*. Cambridge University Press, Cambridge, 1993.

[ND96] G. Nicolis and D. Daems. Nonequilibrium thermodynamics of dynamical systems. *Journal of Physical Chemistry*, **100**:19187, 1996.

[Nic95] G. Nicolis. *Introduction to Nonlinear Science*. Cambridge University Press, Cambridge, 1995.

[Ott92] E. Ott. *Chaos in Dynamical Systems*. Cambridge University Press, Cambridge, 1992.

[Par88] W. Parry. Equilibrium states and weighted uniform distribution of closed orbits. In J. Alexander, editor, *Dynamical Systems*, Lecture Notes in Mathematics. Springer-Verlag, New York, 1988.

[Pen89] R. Penrose. *The Emperor's New Mind*. Oxford University Press, Oxford, 1989.

[Pes77] Ya. B. Pesin. Geodesic flows in closed Riemannian manifolds without focal points. *Russian Academy of Sciences, Izvestiya*, **71**:1252,1447, 1977.

[Pes97] Ya. B. Pesin. *Dimension Theory of Dynamical Systems.* University of Chicago Press, Chicago, 1997.

[Pet91] K. Peterson. *Ergodic Theory.* Cambridge University Press, Cambridge, 1991.

[Poi93] H. Poincaré. *New Methods of Celestial Mechanics.* American Institute of Physics, New York, 1993.

[Pol93] M. Pollicott. *Lectures on Ergodic Theory and Pesin Theory on Compact Manifolds.* Number 180 in London Mathematical Society Lecture Note Series. Cambridge University Press, Cambridge, 1993.

[PR75] Y. Pomeau and P. Resibois. Time dependent correlation functions and mode-mode coupling theories. *Physics Reports*, **19C**:63, 1975.

[PY79] G. Pianigiani and J. A. Yorke. Expanding maps on sets which are almost invariant: Decay and chaos. *Transactions of the American Mathematical Society*, **252**:351, 1979.

[PY98] M. Pollicott and M. Yuri. *Dynamical Systems and Ergodic Theory.* Number 40 in London Mathematical Society Student Texts. Cambridge University Press, 1998.

[Rad96] G. Radons. Suppression of chaotic diffusion by quenched disorder. *Physical Review Letters*, **77**:4748, 1996.

[Rad99] G. Radons. Disorder phenomena in chaotic systems. *Advances in Solid State Physics*, **38**:439, 1999.

[RC98] L. Rondoni and E. G. D. Cohen. Orbital measures in nonequilibrium statistical mechanics: the Onsager relations. *Nonlinearity*, **11**:1395, 1998.

[RdL77] P. Resibois and M. de Leener. *Classical Kinetic Theory of Fluids.* John Wiley, New York, 1977.

[RE85] D. Ruelle and J.-P. Eckmann. Ergodic theory of chaos and strange attractors. *Reviews of Modern Physics*, **57**:617, 1985.

[Rei98] L. Reichl. *A Modern Course in Statistical Physics,* 2nd edition. John Wiley, New York, 1998.

[RM97] L. Rondoni and G.P. Morriss. Applications of periodic orbit theory to n-particle systems. *Journal of Statistical Physics*, **86**:991, 1997.

[Rob99] C. Robinson. *Dynamical Systems: Stability, Symbolic Dynamics, and Chaos.* CRC Press, New York, 1999.

[RS86] D. Ruelle and Ya. G. Sinai. From dynamical systems to statistical mechanics and back. *Physica A*, **140**:1, 1986.

[Rue76] D. Ruelle. A measure associated with Axiom-A attractors. *American Journal of Mathematics*, **98**:619, 1976.

[Rue78] D. Ruelle. *Thermodynamic Formalism.* Addison-Wesley Publishing Co., New York, 1978.

[Rue86a] D. Ruelle. Locating resonances for Axiom-A dynamical systems. *Journal of Statistical Physics*, **44**:281, 1986.

[Rue86b] D. Ruelle. Resonances of chaotic dynamical systems. *Physical Review Letters*, **56**:405, 1986.

[Rue89] D. Ruelle. *Elements of Differentiable Dynamics and Bifurcation Theory.* Academic Press, New York, 1989.

[Rue91] D. Ruelle. *Chance and Chaos.* Princeton University Press, Princeton, 1991.

[Rue96a] D. Ruelle. Positivity of entropy production in non-equilibrium statistical mechanics. *Journal of Statistical Physics*, **85**:1, 1996.

[Rue96b] D. Ruelle. Random smooth dynamical systems. *Lecture Notes*, Rutgers University, 1996.

[Rue97a] D. Ruelle. Differentiation of SRB states. *Communications in Mathematical Physics*, **187**:227, 1997.

[Rue97b] D. Ruelle. Positive entropy production in the presence of a random thermostat. *Journal of Statistical Physics*, **86**:395, 1997.

[Rue98] D. Ruelle. Smooth dynamics and new theoretical ideas in nonequilibrium statistical mechanics. *Lecture Notes*, Rutgers University, 1998.

[Sas96] S. Sastry. Lyapunov spectra, instantaneous normal mode spectra, and relaxation in the Lennard-Jones liquid. *Physical Review Letters*, **76**:3738, 1996.

[Sch89] H. G. Schuster. *Deterministic Chaos,* 2nd edition. VCH Verlagsgesellschaft mbH, Weinheim, 1989.

[Sch97] L. Schulman. *Time's Arrows and Quantum Measurement.* Cambridge University Press, Cambridge, 1997.

[Sim97] N. Simányi. Ergodicity of hard spheres in a box. *to be published*, 1997.

[Sin72] Ya. G. Sinai. Gibbs measures in ergodic theory. *Russian Mathematical Surveys*, **27**:21, 1972.

[Sin89] Ya. G. Sinai, editor. *Dynamical Systems, II*, volume II of *Encyclopedia of Mathematical Sciences.* Springer-Verlag, New York, 1989.

[Sin91] Ya. G. Sinai, editor. *Dynamical Systems, A Collection of Papers.* World Scientific, Singapore, 1991.

[Spo91] H. Spohn. *Large Scale Dynamics of Interacting Particles.* Springer-Verlag, New York, 1991.

[SS96] N. Simányi and D. Szász. Hard ball systems are fully hyperbolic. *to be published*, 1996.

[Szá93] D. Szász. Ergodicity of classical billiard balls. *Physica A*, **194**:86, 1993.

[Tab95] S. Tabachnikov. *Billiards. Panoramas et Synthéses.* Société Mathématique de France, Paris, 1995.

[Tél90] T. Tél. Transient chaos. In Hao Bai-lin, editor, *Directions in Chaos*, volume 3. World Scientific, Singapore, 1990.

[TG95] S. Tasaki and P. Gaspard. Fick's Law and fractality of non-equilibrium stationary states in a reversible multibaker map. *Journal of Statistical Physics*, **81**:935, 1995.

[TGD98] S. Tasaki, T. Gilbert, and J. R. Dorfman. An analytic construction of the SRB measures for baker-type maps. *Chaos*, **8**:424, 1998.

[TGN98] T. Tél, P. Gaspard, and G. Nicolis, editors. *Chaos Focus Issue: Chaos and Irreversibility*, volume **8**,2, 1998.

[TKS92] M. Toda, R. Kubo, and N. Saito. *Statistical Physics,* 2nd edition, volume I. Springer-Verlag, New York, 1992.

[TVB96] T. Tél, J. Vollmer, and W. Breymann. Transient chaos: the origin of transport in driven systems. *Europhysics Letters*, **35**:659, 1996.

[UdBL+nt] G. E. Uhlenbeck, J. de Boer, J. L. Lebowitz, et al., editors. *Studies in Statistical Mechanics.* North Holland Publishing Co., Amsterdam, 1962–present.

[UF63] G. E. Uhlenbeck and G. W. Ford. *Lectures in Statistical Mechanics.* American Mathematical Society, Providence, 1963.

[Vai92] S. Vaienti. Ergodic properties of the discontinuous sawtooth map. *Journal of Statistical Physics*, page 1257, 1992.

[Van92] W. N. Vance. Unstable periodic orbits and transport properties of non-equilibrium steady states. *Physical Review Letters*, **69**:1356, 1992.

[vBD95] H. van Beijeren and J. R. Dorfman. Lyapunov exponents and Kolmogorov-Sinai entropy for the Lorentz gas at low densities. *Physical Review Letters*, **74**:4412, 1995.

[vBD96] H. van Beijeren and J. R. Dorfman. Lyapunov exponents and Kolmogorov–Sinai entropy for the Lorentz gas at low densities: Erratum. *Physical Review Letters*, **76**:3238, 1996.

[vBDC+96] H. van Beijeren, J. R. Dorfman, E. G. D. Cohen, H. A. Posch, and Ch. Dellago. Lyapunov exponents from kinetic theory for a dilute, field driven Lorentz gas. *Physical Review Letters*, **77**:1974, 1996.

[vBDPD97] H. van Beijeren, J. R. Dorfman, H. A. Posch, and Ch. Dellago. Kolmogorov–Sinai entropy for dilute gases in equilibrium. *Physical Review E*, **56**:5272, 1997.

[vBLD98] H. van Beijeren, A. Latz, and J. R. Dorfman. Chaotic properties of dilute two- and three-dimensional random Lorentz gases: equilibrium theory. *Physical Review E*, **57**:1, 1998.

[vK71] N. G. van Kampen. The case against linear response theory. *Physica Norvegica*, 5:279, 1971.

[VTB97] J. Vollmer, T. Tél, and W. Breymann. Equivalence of irreversible entropy production in driven systems: An elementary chaotic map approach. *Physical Review Letters*, **79**:2759, 1997.

[VTB98] J. Vollmer, T. Tél, and W. Breymann. Entropy balance in the presence of drift and diffusion currents: an elementary chaotic map approach. *Physical Review E*, **58**:1672, 1998.

[vV90] G. A. van Velzen. *Lorentz Lattice Gases.* Ph. D. thesis, University of Utrecht, Utrecht, 1990.

[vVE87] G. A. van Velzen and M. H. Ernst. Bond percolation in two and three dimensions: numerical evaluation of time-dependent transport properties. *Journal of Statistical Physics*, **48**:677, 1987.

[vVED88] G. A. van Velzen, M. H. Ernst, and J. Dufty. Crossover dynamics in bond disordered lattices. *Physica A*, **154**:34, 1988.

[vZvB98] R. van Zon and H. van Beijeren. Mean-field theory for the largest Lyapunov exponent for a dilute gas. *Physical Review Letters*, **80**:2035, 1998.

[Wal81] P. Walters. *Ergodic Theory.* Springer-Verlag, New York, 1981.

[Wan87] G. Wannier. *Statistical Mechanics.* Dover Publishing Co., New York, 1987.

[Wax54] N. Wax, editor. *Selected Papers in Noise and Stochastic Processes.* Dover Publishing Co., New York, 1954.

[Whi88] E. T. Whittaker. *A Treatise on the Analytical Dynamics of Particles and Rigid Bodies.* Cambridge University Press, 1988.

[WL98] M. Wojtkowski and C. Liverani. Conformally symplectic dynamics and symmetry of the Lyapunov spectrum. *Communications in Mathematical Physics*, **194**:47, 1998.

[Woj85] M. Wojtkowski. Invariant families of cones and Lyapunov exponents. *Ergodic Theory and Dynamical Systems*, **5**:145, 1985.

[WW83] D. Wilkensen and J. F. Willemsen. Invasion percolation: A new form of percolation theory. *Journal of Physics A: Math. Gen*, **16**:3365, 1983.

[You99] L.-S. Young. Recurrence times and rates of mixing. *Israel Journal of Mathematics*, 1999. to appear.

Index

Printed in the United States
By Bookmasters